Deregulation
of
Cable Television

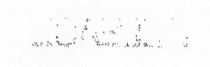

Deregulation
of
Cable Television

Edited by Paul W. MacAvoy

Ford Administration Papers on
Regulatory Reform

HE
8700.7
.C6
D47

American Enterprise Institute for Public Policy Research
Washington, D.C.

Paul J. MacAvoy is professor of economics at Yale University and an adjunct scholar at the American Enterprise Institute.

Library of Congress Catalog Card No. 77-79871

ISBN 0-8447-3254-0

AEI Studies 156

Printed in the United States of America

CONTENTS

LIST OF TABLES

FOREWORD

Early in 1975, I called for the initiation of a major effort aimed at regulatory reform. Members of my administration, and the Congress, were asked to formulate and accelerate programs to remove anti-competitive restrictions in price and entry regulation, to reduce the paper work and procedural burdens in the regulatory process, and to revise procedures in health, safety and other social regulations to bring the costs of these controls in line with their social benefits.

My requests set in motion agency and department initiatives, and a number of studies, reorganization proposals, and legislative proposals were forthcoming last year. A number of these resulted in productive changes in transportation, retail trade, and safety regulations. Nevertheless, much remained to be done, in part because of the time required to complete the analysis and evaluation of ongoing regulations.

This volume provides one set of the analytical studies on regulatory reform that were still in process at the end of 1976. Necessarily, these studies would have undergone detailed evaluation in the agencies and the White House before becoming part of any final reform program. They do not necessarily represent my policy views at this time, but they do contribute to the analyses that must precede policy making. I look forward to the discussion that these papers will surely stimulate.

GERALD R. FORD

PREFACE

This volume is a compendium of studies of cable television policies that were undertaken by President Gerald R. Ford's Domestic Council Review Group on Regulatory Reform (DCRG) from August 1975 through March 1976. Cable television is one of those regulatory problem areas which official and unofficial Washington perennially finds time to study but also finds very difficult to do anything about. In fact, this relatively minor industry has been intensively studied by the Congress, the Federal Communications Commission, various executive branch agencies, and prestigious private organizations including the Committee for Economic Development and the Sloan Commission.

The Ford administration's review of cable television regulation was different from previous studies. First, it was undertaken by analysts who with few exceptions were not experts in telecommunications, much less cable television. Second, the review from the outset was focused on only a few of the economic restrictions that had been imposed by the FCC on cable television, in order to develop a limited but effective program of reform of regulation in this industry.

Previous cable television studies for the most part concluded with the publication of reports recommending various changes in FCC rules. Except in the case of the 1968 Rostow Task Force on Telecommunications Policy, the staff analyses and other materials relating to these reports were not generally published. In contrast, this volume consists almost entirely of supporting materials, although they have been organized in such a fashion as to provide a fairly comprehensive view of the DCRG's cable proposals. Some of these materials, or excerpts from them, have been published previously, but most of them have not. Rather than building to a conclusion, they start from one— that rules on signal use over the cable should in good part be eliminated. They deal with the issue of whether there is sufficient documentary evidence on the effects on consumers to support that policy conclusion.

There are essentially four parts to this report. First, there is introductory material that describes the cable television issues and the DCRG's conclusions about those issues. Second, a memorandum which I prepared in April 1976 arguing against certain proposals is reprinted in full. Third, a rebuttal to my memorandum prepared by recognized authorities in cable regulation economics is reprinted. This is updated by a review of recent policy changes in the last chapter. Finally, we have included as appendixes to this report two of the options papers that were submitted to the DCRG, together with copies of the draft legislation that was prepared by the Office of Telecommunications Policy and the staff of the Antitrust Division of the Department of Justice.

It is axiomatic that those who cannot learn from the past are condemned to repeat it, and that repetition is especially likely when the results of past effort are simply not made available. These studies point clearly in one direction for reform in regulation, that is, towards reforming FCC restrictions on cable use of broadcast signals. Cable associations and consumer groups will continue to push against these restrictions. The studies here point to the work necessary to make their push successful.

PAUL W. MACAVOY
Yale University
May 1977

1

INTRODUCTION AND GENERAL BACKGROUND

Kenneth Robinson

Television, and especially commercial network television, is a familiar and important aspect of modern life. In an average year, television viewers will consume some 142 billion hours of broadcast time, and they will pay indirectly some two cents per hour per household for the opportunity. About 97 percent of all American households have at least one television set, and that average American household will watch television about seven hours a day, seven days a week, fifty-two weeks each year. As one observer noted:

> Television viewing has unquestionably become the leading national pastime, possibly ranking second only to sleep as a national activity. Furthermore, we are often reading about television when we are not watching it. For example, *TV Guide* is our second leading magazine, with only 1 percent less circulation than *Reader's Digest*. . . . A browse through the magazine titles at virtually any newsstand will reveal the inroads that television-gossip magazines have made into a field formerly dominated by movie-gossip magazines.[1]

This is not a phenomenon that the authors of the 1934 Communications Act [2] particularly anticipated in any detail. Writing of television in the late 1930s, Senator Clarence Dill, considered in some ways the father of federal broadcast regulation, wrote:

Kenneth Robinson was formerly with the Office of Telecommunications Policy and is currently employed in the Antitrust Division of the Department of Justice. The views expressed should not be interpreted to be those of the department or of other federal agencies.

[1] D. Bruce Pearson, "Cable: The Thread by Which Television Competition Hangs," *Rutgers Law Review*, vol. 27 (1974), p. 800.

[2] 47 U.S.C., section 151 (1970).

Members of the Commission and employees have attended many demonstrations of television. Television is the new use of radio that may become second only to sound broadcasting as a popular medium of entertainment and information. It will no doubt present many new problems in regulation.[3]

The shape of American television broadcasting today was largely determined by decisions of the Federal Communications Commission reached during the late 1940s. The first decision that the FCC made was concerned with how much of the available radio frequency spectrum should be allocated to television, a matter that was first considered formally in 1941 and was made final about ten years later.[4] Under the National Table of Frequency Allocations,[5] broadcasting and especially television broadcasting was accorded preeminence by the FCC. The commission allocated to broadcasting about 422 megahertz (MHz) of prime radio frequency spectrum space—about 45 percent of the highly desirable frequency band 30 to 960 MHz. About 64 percent of the space in that band was allocated exclusively to non-government use.[6]

The second decision that the FCC made was how to apportion or assign channels within the frequency bands allocated to television. In a ruling finally issued in 1952, the FCC made two important decisions in this regard. To begin with, the commission rejected the proposal of the then-existing Dumont Television Network that the American television system be based upon a relatively small number of very powerful stations, so that some television service would be available to virtually everyone. Instead, the commission opted for a plan based on "localism," whereby there would be, potentially, a large number of local television stations, each serving as a "mouthpiece for its community." Moreover, the FCC also chose to develop

---

[3] Clarence C. Dill, *Radio Law: Practice and Procedure* (Washington, D.C.: National Law Book Co., 1938).

[4] See "Telecommunications—A Program for Progress," Report by the President's Communications Policy Board (Washington, D.C., 1951), pp. 43-46.

[5] The National Table of Frequency Allocations is composed of the U.S. Government Table of Frequency Allocations and the FCC Table of Frequency Allocations (Part 2 of the FCC Rules and Regulations). This resource allocations scheme is developed jointly by the FCC and the President (or his delegate) pursuant to their respective authorities under sections 301 and 305 of the 1934 Communications Act. See Bendix Aviation Corp. v. United States, 272 F.2d 533 (D.C. Cir. 1960), certiorari denied, 361 U.S. 965 (1960).

[6] See *The Radio Frequency Spectrum: United States Use and Management* (Washington, D.C.: Office of Telecommunications Policy, 1973), p. D-36.

American television broadcasting making use of both the very-high-frequency (VHF) and the ultrahigh-frequency (UHF) bands. The third decision that the FCC made was to continue policies that had developed in the case of radio broadcasting whereby the service would be advertiser-supported but there would be in theory sufficient stations competing with each other in any locale to assure "free competition in the marketplace of ideas" and to make government regulation of either the cost of components or the prices charged for airtime unnecessary. Even before the FCC was established, federal policy had been to rely upon competition to assure the reasonableness of prices charged to and by commercial broadcasters.[7] The traditional assumption, too, had been that there was a "public service easement tied to the initial gift" of a frequency license, and that competition among stations was a surer way of achieving diversity than was detailed federal content regulation.[8]

The consequences of these early FCC decisions became apparent in the following years. Because of the economics of advertiser-supported television, the number of television stations originally envisioned failed to materialize. The 1952 plan had called for some 2,000 stations serving some 1,300 communities; fewer than half the stations planned actually went on the air, and indeed in communities where VHF and UHF stations competed, a large number of UHF stations failed.[9] In rural areas the population base proved too small to support many stations; in urban areas insufficient stations were able to be assigned. Because the policy in favor of "localism" meant that signal coverage areas had to be reduced, a significant number of U.S. households found themselves without any television service. For example, the 70 percent of the U.S. population that lived in urbanized areas averaged five or more channels of television available over-the-

---

[7] See Sta-Shine Products Co. v. Station WGGB of Freeport, N.Y., Carman, Proprietor, 188 I.C.C. 271, 278 (1932).

[8] See generally U.S. Congress, Senate, *Hearings on S. 1333 before the Senate Committee on Interstate and Foreign Commerce*, 80th Congress, 1st session, 1947, pp. 201, 304; R. H. Coase, "The Federal Communications Commission," *Journal of Law and Economics*, vol. 2 (1959); Harry Kalven, Jr., "Broadcasting and the First Amendment," *Journal of Law and Economics*, vol. 10 (1967), p. 31; Dallas W. Smythe, "Facing Facts about the Broadcast Business," *University of Chicago Law Review*, vol. 20 (1952), p. 104; Glen O. Robinson, "The FCC and the First Amendment: Observations on 40 Years of Radio and Television Regulation," *Minnesota Law Review*, vol. 52 (1968), p. 70.

[9] See Stanley M. Besen and Paul J. Hanley, "Market Size, VHF Allocations and the Viability of Television Stations," *Journal of Industrial Economics*, vol. 24 (September 1975), p. 41; Joel B. Dirlam and Alfred E. Kahn, "The Merits of Reserving the Cost-Savings from Domestic Communications Satellites for Support of Educational Television," *Yale Law Journal*, vol. 77 (1968), p. 498.

3

air. By contrast, nearly 6 million U.S. households found themselves without any television service for all practical purposes.[10]

Economically, U.S. television developed as an "interlocking grid of shared monopoly markets" at the local level, with most of their programming fed to these local stations by a handful of national television networks.[11] Instead of "localism," American television became, for the most part, a national broadcasting service; some 65 percent of the typical network-affiliated station's programming and more than 80 percent of its profits derived from rebroadcasting locally network "feed" that was imported from New York and other urban centers.

Cable Television Development

Cable television generally refers to the use of coaxial cable to deliver high-clarity television-grade signals directly into subscribers' homes. A super-high antenna typically is erected; over-the-air television signals are received and amplified, and they are then retransmitted via coaxial cable or "wires" to subscribers who pay a monthly charge for the service provided by the community antenna television (CATV) system. Cable television systems vary widely in their degree of technical sophistication; some merely serve as master antenna systems, while others make available a large number of additional channels that can be programmed for other purposes.[12]

Cable television systems developed as a direct result of consumer demand for clearer television reception and more viewing options than were made available under the frequency-allocation plan of 1952 men-

---

[10] As late as 1974, a study determined that 1.2 million U.S. households were not receiving adequate television service on any channel, whether from originating broadcasters, translators, or CATV systems. About 2.4 million households can receive only one channel; 2.2 million only two channels. Denver Research Institute, *Broadband Communications in Rural Areas: National Cost Estimates and Case Study* (Denver: University of Denver, 1974).

[11] See, for example, Alfred E. Kahn, *The Economics of Regulation*, vol. 2, *Institutional Issues* (New York: John Wiley and Sons, 1971), p. 40; Comments of the U.S. Department of Justice in FCC Docket No. 20418 (VHF Drop-Ins), December 1975, reprinted in U.S. Congress, Senate, *Hearings on S. 2028 before the Antitrust and Monopoly Subcommittee of the Judiciary Committee*, 94th Congress, 1st session, 1976, p. 974.

[12] See generally Committee for Economic Development, Research and Policy Committee, *Broadcasting and Cable Television: Policies for Diversity and Change* (New York: Committee for Economic Development, 1975); Sloan Commission on Cable Communications, *On the Cable: The Television of Abundance* (1971); Anne W. Branscomb, "The Cable Fable: Will It Come True?" *Journal of Communications*, vol. 25 (1975), pp. 44-56.

4

tioned earlier.[13] Much of the initial growth of what came to be known as the cable television industry occurred in areas that were completely unserved by any broadcast television. Even today, nearly half of the total 3,500 CATV systems (and about 40 percent of subscribers) are located in areas which are thirty-five miles or farther from the nearest television broadcasting station.

The growth of cable television has not been "explosive," assertions to the contrary notwithstanding.[14] The first cable television system began service to Mahanoy City, Pennsylvania, in 1948. Today, nearly thirty years later, about 11 million of the nation's 90 million television households, or about 15 percent, subscribe to cable television services. By comparison, broadcast television went from a rounding error in 1947 to about 50 percent penetration of the nation's homes in five years, and in ten years it had reached, essentially, 100 percent penetration. Tables 1 and 2 contain some basic statistics about the cable television industry, as supplied by the National Cable Television Association (NCTA), the industry's principal spokesman.

Cable Television Regulation

Federal regulation of cable television services dates essentially from the early 1960s and has grown in scope and intensity in direct proportion to the threat, perceived or actual, which cable services pose to established over-the-air broadcasters. The statutory basis for regulation by the Federal Communications Commission is section 303(g) of the 1934 Communications Act, which admonishes the commission "to promote the wider and more effective use of radio."[15] In *Southwestern Cable*, the Supreme Court affirmed the authority of the FCC to impose restrictions on the number of "distant signals" a cable television system could "import" and make available to its subscribers. Distant-signal importation activities, as the commission reasoned, had to be regulated, lest they "undermine" the system of frequency allocations previously established.[16]

The Court ruled that the commission had the authority under the 1934 act to adopt rules and regulations as "reasonably ancillary" to its

---

[13] U.S. Congress, House, Subcommittee on Communications of the Committee on Interstate and Foreign Commerce, *Staff Report on Cable Television: Promise Versus Regulatory Performance*, 94th Congress, 2d session, 1976.

[14] See, for example, United States v. Midwest Video Corp., 406 U.S. 649, 675-76 (1972).

[15] 47 U.S.C., section 303(g)(1970).

[16] United States v. Southwestern Cable Co., 392 U.S. 157, 178 (1968).

5

Table 1

BASIC STATISTICS ABOUT THE CABLE TELEVISION
INDUSTRY, NOVEMBER 1975

Size of Industry
Number of CATV systems: approximately 3,450
Number of communities served: 7,700
Number of homes served: 10.8 million
Miles of plant in place: approximately 190,000 miles
Penetration (homes served as a ratio of homes passed by cable):
54 percent
Saturation (total CATV subscribers as a ratio of total TV homes):
15.3 percent
Estimated 1975 revenues: $770 million
Employment: 25,000 people

Capital Investment
Total capital investment: approximately $1 billion
Construction costs: range from $3,500 per mile in rural areas to
$10,000 per mile in urban areas. Where underground conduit
must be laid costs can exceed $90,000 per mile

Subscriber Fees
Typical one-time installation fee: $10–15
National average monthly fee: $6
Range of monthly fees in typical larger market systems (built since
1972): $7–10

Services
Retransmission of local broadcast signals
Importation of independent TV stations from nearby cities (subject
to FCC regulations)
Origination of local programming
News ticker, stock reports, weather reports
Public access channels
Pay-cable channels to approximately 700,000 subscribers

Location
All 50 states plus the Virgin Islands, Puerto Rico, and Guam

Source: National Cable Television Association, November 12, 1975.

regulation of television broadcasting. The FCC promptly initiated a
series of rule-making proceedings examining into the kinds of regulations that should be adopted. In 1972, regulations requiring cable
television systems to originate certain kinds of programming were
affirmed by the Supreme Court as reasonably ancillary to the commission's general obligation to promote diversity.[17]

[17] See United States v. Midwest Video Corp., 406 U.S. 649, 662-63 (1972).

Table 2
STATE STATISTICS ABOUT THE CABLE TELEVISION INDUSTRY, NOVEMBER 1975

	Systems	Communities	Subscribers
Alabama	81	120	189,938
Alaska	14	18	9,284
Arizona	37	64	66,213
Arkansas	75	98	109,532
California	286	709	1,472,921
Colorado	39	84	78,296
Connecticut	11	29	64,130
Delaware	8	29	60,376
Florida	112	296	456,195
Georgia	74	140	232,422
Hawaii	10	54	36,556
Idaho	45	76	56,768
Illinois	77	175	307,084
Indiana	66	103	207,102
Iowa	41	53	77,704
Kansas	84	101	150,331
Kentucky	108	217	149,641
Louisiana	37	61	93,403
Maine	32	59	54,085
Maryland	27	74	91,952
Massachusetts	42	76	160,085
Michigan	71	227	256,299
Minnesota	84	117	129,918
Mississippi	62	95	150,675
Missouri	65	94	127,353
Montana	32	51	88,312
Nebraska	46	50	58,311
Nevada	6	10	28,405
New Hampshire	38	76	76,783
New Jersey	33	147	204,711
New Mexico	31	54	91,010
New York	165	562	789,686
North Carolina	44	81	140,562
North Dakota	12	15	25,496
Ohio	148	363	530,054
Oklahoma	79	90	153,760

Table 2 (continued)

	Systems	Communities	Subscribers
Oregon	92	187	167,853
Pennsylvania	301	1,398	1,047,960
Rhode Island	1	1	3,091
South Carolina	38	59	80,994
South Dakota	17	22	35,113
Tennessee	59	79	112,206
Texas	234	309	651,831
Utah	6	9	6,845
Vermont	35	90	53,398
Virginia	60	124	139,040
Washington	100	215	252,268
West Virginia	142	368	242,268
Wisconsin	70	129	143,086
Wyoming	25	39	61,270
Guam	1	1	9,400
Puerto Rico	1	1	12,000
Virgin Islands	1	1	975

Source: National Cable Television Association, November 12, 1975.

Collateral decisions in the copyright field made the FCC's exercise of this ancillary jurisdiction considerably more complicated. In the *Fortnightly* case, the Court had ruled that the reception and retransmission of broadcast television signals by cable systems did not constitute a "public performance for profit" within the narrow meaning of the 1909 Copyright Act.[18] Established television interests argued that, because of these rulings, cable television constituted "unfair competition" for broadcasters who were obliged to pay copyright royalties.

By 1974, the FCC in response to broadcasters' arguments had fashioned an enormously complex regulatory system to control cable television. Complicated regulations governed both the number of distant signals a cable television system might import and the particular nonlocal areas from which they might be imported. Additional regulations governed what kinds of programs might be so aired; certain kinds of nonnetwork, syndicated programs were required to be

---

[18] Fortnightly Corp. v. United Artists Television Corp., 392 U.S. 390, 400 (1968); TelePrompTer Corp. v. CBS, Inc., 415 U.S. 394, 405 (1974).

"blacked out" as a surrogate for copyright liability. The rationale behind these complicated rules was the need to guard against "audience siphoning." The commission reasoned that the number of viewing options available to cable television subscribers must be restricted to avoid "siphoning away" the audience basis for local television stations and, as a consequence, the profits necessary in theory to sustain various kinds of public service programming required by the FCC.

To supplement these distant-signal regulations, the FCC adopted so-called pay-cable rules. In distinction to the distant-signal rationale ("audience siphoning"), the pay-cable rules were based upon the need to guard against cable systems "bidding away" the most popular kinds of "free television" fare, by selling those programs to subscribers directly on a per-channel or per-program charge basis. As ultimately refined, the pay-cable rules essentially barred cable television systems from offering any movies that were more than three years old on a pay-cable basis. Sports broadcasts were protected from siphoning by way of a complex "highwater mark rule," which allowed cable systems to air only half the number of games of an event that local television stations chose not to air, subject to various bizarre qualifications.

Overall, the commission proceeded to adopt a large number of regulations with respect to cable television operations that had only the most remote or tangential relationship to the perceived threat to the maintenance of "free television." Technical standards were adopted; equal employment and related hiring rules were adopted; the ability of local governments to levy franchise fees was restricted, as was their ability to grant multiyear franchises. Detailed procedures were devised to govern the granting of franchises. Severe limitations were imposed on the origination of advertisements or commercials by cable systems. Cable television originations were made subject to the FCC's "fairness" and "equal time" regulations. The commission initiated so many rule makings with respect to cable television regulation that a special "clarification" proceeding was initiated to clarify the relationship of the other pending proceedings. (As it developed, the so-called clarification provoked such controversy and additional confusion that it was evidently rescinded.)

The Basis for Domestic Council Action

The Domestic Council Review Group on Regulatory Reform (DCRG) was requested to undertake a review and evaluation of cable television regulation by the Office of Telecommunications Policy (OTP) in the Executive Office of the President. In 1971, OTP had itself begun a study into FCC cable television regulation by forming a Cabinet Committee on Cable Television, subsequently renamed to become the Cabinet Committee on Cable Communications.

Meeting irregularly over the next two-and-one-half years, the OTP study group had focused on dangers to First Amendment freedoms it perceived in the steady growth and complexity of FCC cable television regulations. Nearly three years after it was impaneled, the OTP study group released a report generally known as the Whitehead Report, after Clay T. Whitehead, the OTP director. In May 1974, OTP proposed "comprehensive cable television legislation" to implement the various recommendations contained in the Whitehead Report and submitted the draft bill to the Office of Management and Budget in accordance with established clearance procedures.

This first OTP proposal was sharply criticized by the FCC as too restrictive of the "flexibility" which the commission argued was required in order to regulate the fast-growing cable television industry in the "public interest." At the same time, the Department of Justice criticized the OTP proposal for affording the FCC too much flexibility and for failing to address those aspects of the commission's cable regulation it believed were most economically significant, namely the restrictions on distant-signal importation and on pay cable.

Subsequent changes in OTP's basic bill were of a minor nature. OTP declined to include provisions that would relax the FCC's distant-signal and pay-cable regulations on the ground that to do so would prompt opposition by broadcast interests. Recommendations that the implementation of the so-called separations policy be clarified were also not adopted.

The Department of Justice's basic arguments against submission of the legislation proposed by OTP were contained in a lengthly letter, the full text of which is reprinted here in Appendix A. The department's views were summarized, however, when Deputy Assistant Attorney General Donald I. Baker testified before the late Senator Philip A. Hart's (Democrat, Michigan) Antitrust and Monopoly Subcommittee in July 1975:

> The expansive character of the [FCC's] pay cable rules, of course, raises serious public policy implications. As the subcommittee recognizes, it is only since 1968 that the FCC has had *any* clear legal authority with respect to cable television. Yet in seven years, the FCC has expanded its regulation to

cover parties whose relationship with interstate broadcasting is fairly indirect. In view of the breadth of the FCC's cable regulations, and their competitive and social implications, a number of people have called for a congressional solution to the matter and, indeed, a number of alternative courses of action are available to Congress.

One alternative, of course, is to do nothing. A similar result could be obtained by ratifying the status quo through legislation. . . . Another alternative would be to simply deregulate the cable television industry all together [sic]. Such an alternative is not quite so radical a concept as it might seem at first. The case for detailed cable television regulation, while it may be developed in the future, is presently not that compelling. The oft-repeated concern that cable growth could imperil the continued viability of broadcasters serving sparsely populated rural areas seems on closer examination to be less than convincing. For example, as the recent OTP study on "Rural Areas" demonstrated, there are very large areas of the country where there is not *any* broadcast television available to begin with. That study concluded, for instance, that something like 6 million U.S. households—or something like two-thirds the number as now get cable—currently receive two television signals or less. Indeed, the FCC rules do not, contrary to the public image upon which they are based, actually afford much, if any, protection for rural broadcasters at the present time. Under the so-called "Cable Consensus," of 1971–72, the greatest exclusivity protection is afforded stations in the top fifty markets (who arguably, need the protection the least). Somewhat less protection is afforded the "Second 50" stations, and below the top 100, the television stations are so to speak thrown to the "piranhas."

The other traditional concern which has been used to rationalize the need for detailed cable regulation arises from the situation in which the local cable system may be a de facto monopolist. This is, in situations where the only way to receive a satisfactory television signal is to subscribe to the local cable system. Such systems, of course, exist, and having lived with one myself, I can appreciate this particular argument in favor of regulation. However, as I said earlier, the situation is relatively infrequent. Seventy percent of the population lives in areas where five or more television signals are available over the air. The existence of monopoly power in the hands of a cable system in a limited number of areas does not justify applying detailed restrictions upon cable in all markets.[19]

Baker went on in his testimony to set forth a "laundry list" of issues which the Department of Justice had argued should be addressed if the Ford administration chose to tackle the question of cable regulation.

> Assuming . . . that complete deregulation is not considered to be a practical alternative, then the question arises, what kind of legislation would be desirable?
>
> A first requirement of such legislation would be that it reflect a policy of providing competitive opportunity for cable television, along the lines I discussed. . . . Providing competitive opportunity is not the same as promoting cable or any other media as a "chosen instrument" of Congress. The latter policy may impede the development of a future technology. The restrictive practices of today would be repeated to protect the status quo of some future period.
>
> A second major consideration is that any legislation proposed take into account the plain fact that cable television as presently provided is indeed a capital intensive industry. In major urban areas, for example, the cost of constructing a system can range upwards to $800 per subscriber. Common sense dictates that capital intensive industries require a certain degree of stability and profit incentives in order to develop. Delegations of broadly worded regulatory power to the FCC or the imposition by Congress of extraordinary capacity and extra service demands far in advance of conceivable marketplace demand would serve as deterrents to the investment of capital which is a prerequisite for the growth of the cable industry.
>
> A third and related major consideration is that any legislation proposed should be whole unto itself, and specify clearly who *is* to do what as well as who *can* do what. If legislation is required in order to remedy problems brought about by too much administrative "flexibility," then the legislation proposed should not contain numerous provisions that rely for their meaning upon future "reasoned elaboration" by an administrative agency.
>
> A fourth major consideration which the Department believes is essential is that any legislation endeavor to deal with existing as well as any future competitive and regu-

---

[19] U.S. Congress, Senate, Testimony of Donald I. Baker before the Antitrust and Monopoly Subcommittee of the Judiciary Committee, 94th Congress, 1st session, July 1975.

latory difficulties. The competitive difficulties confronting the cable industry today . . . are genuine and not imaginary. It would not be responsible for any proposed legislation to show a blind eye towards these significant problems on the basis that possible future developments—should they ever occur—will make these present day difficulties seem insignificant. Unless the existing restrictions upon the potential of cable are ameliorated, the industry may never attract the capital necessary for it to provide consumers with additional choice. Should cable eventually develop into the dominant media, the FCC could be given adequate regulatory powers limited to the public interest problems actually experienced. However, an infant industry should not be fettered with chains designed to handle a giant.

Finally, any legislative proposal should take into account the competitive objections that we have raised over the years with respect to cross-ownership of media. Presently, it is true that cable television and television stations are not so vigorous competitors in local media markets as perhaps we would like. One reason why this is the case is that the FCC has imposed very strict constraints upon the amount of advertising a cable system may show as well as strict constraints on the type of programming it can show. Nevertheless, the entire premise of broadcast regulation—and the profit base of the industry—rests upon the scarcity of frequencies available for assignment under the FCC's regulations. The whole concept of scarcity is potentially threatened by cable television, because it offers a possible regime of abundance—of more choice at incremental costs. In view of these potential developments, sanctioning the ownership of a cable system by a television station serving the same community involves very real competitive risks. Since the FCC rules, and competitive considerations, make it possible for broadcasters to invest in cable systems in every market in the country, except for the maximum of seven in which they are allowed to own television stations, we see very little to be gained and much to be lost by allowing cable television cross-ownership in the same community.[20]

Initial Domestic Council Actions

Following the Baker testimony, OTP revised its cable television legislative proposal, and in late August 1975 the office submitted this bill with an explanation to the DCRG. In the interim, the staff of the

[20] Ibid.

Department of Justice had reworked earlier OTP draft bills into a "Comprehensive Cable Communications Act of 1976," which was nothing if not comprehensive. Both pieces of proposed legislation and related materials are reprinted here as Appendixes B and C.

As a first step toward developing an administration position, the DCRG undertook to solicit the views of all interested nongovernment groups. Over the next few months the DCRG and its membership met with more than 200 people to discuss cable television legislation. Formal meetings of the DCRG and interested parties were held in the Roosevelt Room of the White House. A great deal of supplementary material was also submitted. Table 3 lists the formal meetings on cable legislation that were undertaken.

At each meeting, a formal agenda was circulated in an effort to focus the discussion on those issues that might possibly be addressed in any proposed administration legislation. The agenda used was essentially the same at all formal meetings. The following agenda was submitted to the DCRG members in conjunction with the formal meeting with the boards of directors and other officials of the National Cable Television Association and the Community Antenna Television Association on October 9, 1975:

> The following are the points of interest for discussion in the above meeting:
>
> 1. In what substantive areas must cable be regulated to protect the public interest and which aspects of cable operations can be left safely to a marketplace determination?
>
> 2. To what extent, if any, should the Government limit intermedia competition between the cable and broadcast television industries; what impact would deregulation of signal transmission or pay television have on the public's reception of television programming?
>
> 3. Is cable subject to inconsistent and duplicative Federal, State, and local regulation; if so, does such regulation adversely affect interstate commerce or otherwise impose costs on the public; how can this problem be alleviated?
>
> 4. What is the proper allocation of regulatory authority over cable, between Federal and non-Federal interests?
>
> 5. Discuss cable problems in the following areas:
>   (a) Access to poles and ducts.
>   (b) Discriminatory taxes.
>   (c) Licensing fees.[21]

---

[21] Memorandum from Lynn May to the DCRG Members on "Agenda for the DCRG/Cable Industry Meeting, Thursday, October 9, 1975," The White House, October 8, 1975.

Table 3

FORMAL MEETINGS WITH THE DCRG ON CABLE LEGISLATION, SEPTEMBER 1975–FEBRUARY 1976

Parties	Date	Place
Congressional		
House Communications Subcommittee	Wednesday, September 24, 1975	Rayburn House Office Building
Senate Communications Subcommittee	Monday, September 29, 1975	Dirksen Senate Office Building
Senate Antitrust and Monopoly Subcommittee	Friday, October 24, 1975	New Senate Office Building Annex
Industry		
Cable industry (National Cable Television Association, Community Antenna Television Association)	Thursday, October 9, 1975	White House
Television/broadcasting (National Association of Broadcasters, Association of Maximum Service Telecasters, networks, National Translator Association, Corporation for Public Broadcasting)	Thursday, October 16, 1975	Old Executive Office Building
Motion picture/program production industries (Motion Picture Association of America, major firms, et cetera)	Monday, October 20, 1975 Wednesday, February 11, 1976	White House Old Executive Office Building
Professional sports (all leagues except National Football League)	Thursday, October 23, 1975 Friday, October 31, 1975	White House Old Executive Office Building
Theatre owners (National Association of Theatre Owners, et cetera)	Friday, October 24, 1975	White House
Public interest groups, state and local governments	Tuesday, October 28, 1975	White House
Academia	Friday, November 21, 1975	1800 G Street

As this agenda demonstrates, the topics for discussion at each of the formal meetings of the DCRG ranged across all aspects of cable operation. At the outset, all of the topics raised in the legislation proposed by OTP were considered apt topics for discussion. It soon became relatively clear, however, that the two features of the FCC's cable regulations that were considered to be most economically significant were the restrictions on distant-signal importation and the restrictions on pay-cable programming. Consequently, most of the discussion after the first meeting (with the cable interests) centered on these twin issues.

From time to time, the view was expressed that the DCRG should endeavor to resolve such matters as copyright liability; alternatively, it was proposed that there should be legislation dealing with a broad range of media-related topics, not just cable television issues. The Department of Justice representative pointed out that it was possible, although not necessarily desirable, to expand the scope of the review to include more general media-related topics.

The following memorandum from Deputy Assistant Attorney General Jonathan C. Rose of the Antitrust Division of the Department of Justice set forth for discussion a means in which this expansion of scope might be accomplished:

> Development of a broadly based Administration deregulation initiative in the communications field is possible, and might be helpful in affording another decisional option. A plausible "combination offering" was sketched briefly for Lynn May and yourself last Friday, October 24. It would entail a proposed "de-thicketing" of over-the-air broadcasting's regulation including program and other content constraints. It would also include selective relaxation of the most competitively significant FCC cable television restrictions. These changes would be keyed to each other, and conditioned upon the emergence of more effective competition among media in local or metropolitan markets. Effective competition among local media would diminish the traditional case for detailed regulatory scrutiny of broadcasters; expanded cable television operations could contribute such competition.
>
> Important features of such a proposal would include, one, cutting back those regulatory restraints on cable TV we consider most significant, subject to the "adverse impact" caveats both the OTP and Justice bills now contain. This would essentially mean a lifting of the present pay and distant signal programming restraints in most metropolitan regions. . . .

Two, it would propose extending to 10 years the present 3-year maximum broadcast license term. License terms would be adjustable according to FCC regulations, upon a determination that effective competition warrants such action. This would basically mean 10-year license terms for almost all radio stations, longer license terms for television stations in many markets below 100 ranking—where catv presently has achieved some significant market penetration —and possibly longer license terms for major market television licensees as they may confront more viable cable television competition. . . .

Third, statutory "Fairness" and "Equal Time" requirements and other broadcast program controls would be rescinded, wherever "sufficient" competing channels of electronic communications existed to effectively undermine the scarcity rationale that underlies these controls. This would probably result in lifting such regulations from metropolitan area broadcasters, where the number of media outlets afforded the public is, of course, not limited. To the extent that market telecasters might effectively establish a case for FCC protection from the competitive inroads of cable, the regulation of their business operations would, relatively speaking, increase.[22]

The Second Phase of Domestic Council Activities

Although consideration was given to expanding the scope of the DCRG's cable television regulation review, by November 1975 the focus had shifted back to the original purposes of the undertaking. The focus thereafter was to be cable television policies only, and more specifically, cable policies as they pertained to distant-signal and pay-television activities. The Domestic Council staff had mapped out a course of action in a memorandum about a DCRG meeting on cable reform to Roderick Hills, a presidential adviser, who was subsequently named chairman of the Securities and Exchange Commission.

The discussion centered on whether or not the DCRG should adopt the OTP legislation or opt instead for a simple deregulatory bill dealing with distant signal transmission

---

[22] Memorandum from Jonathan C. Rose, deputy assistant attorney general, Antitrust Division, to Calvin J. Collier, general counsel, Office of Management and Budget, on "Proposed Cable Television Legislation," Department of Justice, October 28, 1975.

and/or pay cable as advocated by Justice. While no conclusion was reached, the group agreed to the following:

1. Jon Rose (Justice) should prepare legislation that would de-regulate distant signal retransmission and separate language to remove FCC controls on pay television.

2. Lynn May would set up meetings for DCRG members to discuss the issue with appropriate Congressional Staff Members and with special interest groups. . . .

3. Following the above, Paul MacAvoy, Stan Morris and Lynn May will write an issue paper for the EPB [Economic Policy Board] providing the following options for Administration cable legislation:

    a. The comprehensive OTP bill.
    b. A bill de-regulating long distant signals and/or pay cable.
    c. Both of the above bills to be submitted separately.[23]

Following the meetings that were scheduled and subsequent meetings among the members of the DCRG, this earlier memorandum was "resurrected" to become the plan for future actions. Proposals suggesting either a broadening of the scope of the review or other decisional options were dropped, and attention was focused on the tasks assigned to each of the DCRG members.

As a first step toward the preparation of a formal issues paper, Paul MacAvoy requested that separate papers be prepared summarizing relevant points about the distant-signal and pay-cable points. One of these "options papers," dealing with pay-cable issues is reprinted as Appendix D of this report. In preparing these separate options papers the staff members responsible were instructed to narrow the effort to cover only the topic assigned; extraneous and additional topics and information were to be excluded in order to avoid further confusing the issues.

While the DCRG was returning to a more narrowly focused effort preliminary to proposing specific, "targeted" draft legislation, however, the commercial broadcasting industry initiated a lobbying campaign to forestall the submission of any administration legislation dealing with cable. Draft letters to be sent to the President and the White House staff personnel were mailed to broadcasters across the country by William Carlisle, vice president for governmental relations of the National Association of Broadcasters. The Carlisle

---

[23] Memorandum from Lynn May to Rod Hills on "DCRG Meeting on Cable Reform," The White House, September 16, 1975.

transmittal letter, copies of which were provided by one recipient to the DCRG, stated as follows:

Cable television regulatory legislation is being formulated simultaneously in three Federal sectors: the White House Domestic Council is seeking information in order to produce an omnibus cable bill for the Administration; the FCC has in preparation an omnibus cable bill; and Mr. Macdonald's House Communications Subcommittee has been actively interviewing all interested parties (the networks, NAB, NCTA, etc.) in order to formulate its own legislative proposal. Thus we are entering a key and vitally important time segment which may well determine the nature of the evolution of cable and, most importantly, pay cable.

We would like each member of the Las Vegas Action Group to prepare *original* letters to be sent to President Ford, to the seven FCC Commissioners, and to the eight members of the House Communications Subcommittee. Also, we want you to phone as many broadcasters as you can in your assigned states to write *similar* letters.

In the foregoing paragraph I emphasized the words "original" and "similar" because we want no duplication of language which would give the impression that writers are simply copying a model. Hence, paraphrasing will be absolutely necessary. I am enclosing two sample letters which were prepared by a member of our anti-siphoning committee and which could serve as the basis for letters to all three Federal sectors. Please feel free to vary the order of priorities and the language to the fullest extent. If you have any policy questions, please call Don Zeifang or me.

In any event, your letters to the White House, the FCC and the House Subcommittee should emphasize your independent belief (which I know we all share) that omnibus cable legislation should express the fundamental policy that, since cable services cannot be made available to all of the people in the United States, they should be supplemental to broadcasting services; encouraged to provide new, innovative services, but not permitted to offer the same programs for pay that are now available free to substantially all of the people.

The extent to which you go into such matters as licensing, signal carriage and exclusivity, other originations, technical standards or FCC enforcement (including forfeitures) will depend upon how long and inclusive you wish your letter to be. Furthermore, with the exception of a very strong anti-siphoning position, NAB policy is not necessarily identical to the points raised in the sample letters. Accordingly, express your own views as you see them.

I particularly emphasize that we want very much for you to forcefully contact other broadcasters to undertake this mission. Let us know how we can help, and please blind copy us for our files! [24]

Following the dispatch of this letter, the commercial broadcasting industry pressures against the submission of cable legislation intensified. As one broadcast industry source was quoted as saying:

It was not just a couple of people doing the lobbying. Contacts were made by a variety of people. We put a snow storm of broadcast pressure on all sides of the White House. Seidman was contacted, Marsh was talked to, Buchen's been talked to, and so has Baroody. Everybody was very, very busy pushing the public interest-quote-unquote.[25]

By January 1976 the strenuous opposition of the broadcasting industry had diminished some of the earlier enthusiasm on the part of the DCRG members for proposing cable legislation of any kind. Masses of detailed information and putative "studies" were submitted to the DCRG and others aimed at demonstrating the destructive potentials of any expansion of cable television activities, especially upon small-market television station operations. Some concern was also voiced that the staff analyses that had been prepared inadequately reflected the legitimate concerns of the television industry and were more advocacy papers aimed at persuading the DCRG to act in a "pro-cable" fashion than anything else. MacAvoy in particular was concerned that there seemed to be no empirical basis to sustain the conclusions advanced by either side of the cable controversy. Since they had recently completed studies of federal transportation and energy policies where such an empirical basis had first been laid to support proposed changes, the staff of the Council of Economic Advisers were consulted.

Staff economists in the Council of Economic Advisers reviewed the various materials which had been submitted to the DCRG with respect to cable regulatory policies. They then reviewed the available literature in the area and determined what studies and analyses had previously been done. Particular attention was given to the question whether there was available credible and sufficient material to be able to refute the arguments of the broadcasting industry that any relaxa-

[24] Letter from William Carlisle, National Association of Broadcasters, to [Addressee deleted as requested], September 29, 1975.
[25] *The Media Report*, July 2, 1976, pp. 3-4.

tion in the FCC rules would trigger catastrophic changes. From these materials MacAvoy prepared a final report in the form of an April 2, 1976, memorandum to the DCRG. This memorandum is reprinted in full in the following chapter.

2

MEMORANDUM ON REGULATORY REFORM IN BROADCASTING

Paul W. MacAvoy

For the last six months a DCRG Working Group has reviewed Federal Communications Commission regulations of television broadcasting. Attention has centered on restrictions on the use by cable television companies of broadcast signals, by nature the most restrictive FCC procedures. These restrictions were examined as part of the reform initiative to remove "anticompetitive regulations."

Cable television is a relatively "new" communications medium compared to over-the-air broadcasting. Cable currently makes extensive use of television signals originated by broadcasters. The FCC, however, limits the number of signals that are not locally produced which cable operators may use. The introduction of new technology invariably erodes the position of established firms, but this effect generally benefits the public in terms of new products and competitive prices. This phenomenon would argue for eliminating FCC rules against signal usage. However, a number of significant objections have been raised about the free use of imported signals by cable. First, since copyright payments are not yet required for imported signals, program producers are denied full value of their product. Second, the imported signals could so fragment the markets of local over-the-air broadcasters that local service would be eliminated or radically reduced in quality. This could result in reduced total service to viewers without cable. Third, unrestricted pay cable might "siphon" programs now available free on over-the-air television. Finally, the FCC's policy of localism may be significantly eroded, with consequent adverse effects on the full dissemination of information at the local level.

This memorandum to the Domestic Council Review Group on Regulatory Reform is dated April 2, 1976.

The DCRG Working Group has examined and extended the research literature on cable and pay television and has solicited further analyses of these issues from the cable and broadcast industries. In our judgment this literature does not fully address the issues, nor has further work on our part or on the industries' part been able to produce definitive forecasts of the effects of cable deregulation on (1) the cable industry, (2) the broadcasters, or (3) consumers.

Some preliminary conclusions have been established. The threat to some existing broadcasters could be excessive—in terms of producing adverse effects on viewers without cable—but the evidence on that is extremely slim. The copyright problem may be dealt with by proposed legislation this term, so that the effects of deregulation without copyright are obscure and perhaps irrelevant. The FCC policy of localism could be eroded in mid-sized markets by cable deregulation. But it is now impossible to place any estimate on the real value of local service for consumers. Further signal importation by pay television could produce more and different programs, although this is more likely with entertainment than with sports programs.

However, more research is necessary before burden of proof for regulatory changes can be borne by these preliminary conclusions. For example, even if mid-sized markets will be most affected by relaxed rules on distant-signal importation, we have not been able to estimate how many markets are involved or how great the impact will be on local broadcasting. The research necessary for policy evaluation is outlined in the following sections of this study. Some of the work is extensive, and both procedures and results are uncertain. No schedule can be set at this time for completing the work. It is to be hoped that interested public service institutions, the industries involved, and the FCC might undertake some of this research.

For each issue we review the existing research results, present preliminary conclusions, and suggest the additional work necessary to extend these conclusions. The review starts by treating the development of FCC policy.[1] Next are analyses of the effects of the reform on cable systems, broadcasters, and consumers. Last of all there are discussions of pay television and educational television.

---

[1] The most recent and comprehensive treatments of television regulations are R. G. Noll, M. J. Peck, and J. J. McGowan, *Economic Aspects of Television Regulation* (Washington, D.C.: The Brookings Institution, 1973); and B. M. Owen, J. Beebe, and W. G. Manning, Jr., *Television Economics* (Lexington, Mass.: Lexington Books, 1974).

FCC Broadcast and Cable Regulation

During the 1950s and 1960s, the FCC issued television broadcasting licenses in order to spread stations geographically across the country. With the goal of putting in place 2,000 stations to serve 1,300 communities, the commission attempted to provide more of the smaller communities with their own sources of news and feature programming.

However, it soon became clear that, for technological and economic reasons, VHF broadcasting stations could not be located in many smaller markets, and that only five or six channels could operate in the largest markets. The FCC, particularly concerned about the lack of local service in small communities, proposed to strike at this problem of limited localism by licensing a large number of UHF stations. Thus, the 1952 allocation plan foresaw the operation of at least one or two stations in each community, and most of these on the UHF band.

The FCC's objective was not simply availability of television throughout the nation; that could have been accomplished using only the VHF range.[2] In fact, alternative frequency allocations could increase the number of homes receiving television. The FCC saw local television stations as "instruments for community enlightenment and cohesion, much like the hometown newspaper of an earlier era."[3]

The 1952 allocation plan has not been successful on its own terms. Television has not become something "like" a community newspaper. Further, because of the relatively low population densities found in much of the country, fewer than half of the planned stations are now broadcasting. Those in population centers enjoy access to fewer channels than they would have had in the absence of the allocation plan, while many of those in outlying areas still have few options and relatively poor quality service.

Cable television was drawn by consumer demand into gaps in the market left by the "visible hand" of these policies of the FCC.[4] CATV first appeared in smaller towns and suburban fringes of cities, where off-the-air reception was either impossible or of low quality. The role of CATV in these areas was to provide better television or simply some television. Somewhat later CATV began to appear in core cities, where several high-quality free channels were available

---

[2] Or, in fact, with a single channel, as was done in the United Kingdom.

[3] Noll, Peck, and McGowan, *Television Regulation*, pp. 58-93 and 100.

[4] Leonard Chazen and Leonard Ross, "Federal Regulation of CATV: The Visible Hand," *Harvard Law Review*, vol. 83 (June 1970).

over-the-air. Improved quality, especially quality for UHF channels, was an important consideration in some large markets—especially New York and Los Angeles. But in some cases cable was able to enter these markets because many consumers were willing to pay for the added diversity that it could provide through imported signals.[5]

CATV in the mid-1960s still served only a very small fraction of the television homes in the United States. Nevertheless, the very rapid growth of cable suggested that it would shortly upset the FCC's 1952 allocation plan. In particular, the growth of cable seemed to threaten the viability of UHF stations, most of which were at best marginally profitable, and to limit severely the possibilities for creating more local stations.[6]

The FCC responded to this situation in 1966 by freezing cable in the largest 100 television markets.[7] The importation of additional distant signals by existing systems and the creation of new systems were prohibited in these markets pending further consideration by the commission.

The freeze was lifted in 1972, but neither the regulation of CATV nor the policy of localism was abandoned. In place of the freeze, the FCC imposed three important restrictions. First, the number of distant signals that could be imported was sharply limited.[8] Three nonnetwork signals at most can be imported in the top fifty markets; two nonnetwork signals at most can be imported in the next fifty; and, if there is a local independent, no nonnetwork signals can be imported into the remaining "smaller markets." Second, the FCC imposed exclusivity rules that require blacking out specific programs in imported signals.[9] The restrictions are strongest in the top fifty markets,

---

[5] The relative importance of these two factors cannot be assessed city-by-city at this time without significant additional documentation which is not yet available.

[6] Such, in essence, was the conclusion of F. M. Fisher, V. E. Ferrall, Jr., et al., "Community Antenna Television Systems and Local Television Station Audience," *Quarterly Journal of Economics*, vol. 60 (May 1966), pp. 227-51. That paper, which was the first important econometric work on CATV, is criticized in the present study.

[7] Cable has made its greatest inroads into smaller and fringe markets, and it is in these markets that local stations are least profitable. Consequently, given the value accorded to localism by the FCC, it is anomalous that the freeze was applied to only the top 100 markets. This point, which is treated further in this study, has never been adequately explained by the FCC.

[8] Until recently, the "leapfrogging rule" required that signals be imported from the closest market. The FCC recently rescinded some of these restrictions. See FCC Report and Order 75-1409, December 19, 1975, p. 38342.

[9] Syndicated programs, as well as those under contract to local stations in general, cannot be imported, at least for an appreciable period of time, in the

where most movies and serials of the networks are blacked out of a distant signal in favor of the local signal. As a result, the distant signal is typically blank half of the time. Third, pay-cable services are limited to providing sports events that are not generally televised over-the-air.[10] FCC rules also essentially exclude from pay cable current motion pictures—in some instances, even movies more than ten years old if they have been televised in the community within the past three years.

Individually, and collectively, these restrictions could have a significant impact on CATV. The FCC's restrictions could well retard the rate of growth of cable and limit the value of CATV to subscribers. If the FCC's restrictions were relaxed, the cable company could disseminate the local stations and perhaps as many as half a dozen additional signals. Furthermore, pay cable would likely offer programming tailored to small audiences. The effects both of the restrictions and of their removal have been the subject of a lengthy debate.

The examination of change in regulation centers on removing these restrictions, as follows:

Distant Signals. The restrictions on use of distant signals would be eliminated altogether, subject to the payment of copyright fees on program use as set out in the bill now before Congress. A possible variant would require cable systems to "affiliate" themselves with a major independent station, bargaining for the right to retransmit its signal. Copyright owners would then receive their compensation by charging the station more.

Pay Programming. Two types of pay legislation have been circulated, the first proposing to eliminate the restrictions outright, subject to an "impact" finding, and the second proposing to include language guaranteeing the public the right to see certain programs on "free" television. In the second, a distinction is made between movies and sports, with the limits placed on sports. Distinguishing between

---

top 100 markets. See Cable Television Regulation Report and Order, 36 FCC 2d 143 (1972). Some relatively minor changes in the exclusivity rules have since been made. The restrictions are less stringent in the second 50 markets, and the exclusivity rules do not apply to smaller markets.

[10] The FCC pay-television rules effectively eliminate sportscasts on a pay basis for sporting events shown in the market within the past five years. The "market" is defined to include the market served by any of the television stations whose signals the FCC requires the CATV system to carry. Under the rules, the FCC requires that certain "fringe area" stations be carried, so the market for the purpose of these rules is very broad.

movies and sports programs in the long run might be justified on the grounds that the supply of the first, in theory, is expandable while the supply of the second is not.[11]

Impact on Cable Television

In this section the effect on CATV of removing regulatory restrictions is analyzed. First, the effect of removing the restrictions on imported distant signals is discussed. Because of a number of problems with presently available research results, it is only possible to conclude qualitatively that CATV would experience faster growth, particularly in mid-sized markets. Then, the effect of changes in the exclusivity rules and copyright arrangements is analyzed. It is found that CATV systems can pay copyright fees of 3 percent to 5 percent and still experience modest growth, particularly if the exclusivity rules are ended as part of the copyright solution. Finally, the effect of these changes on the programming originated by CATV is discussed. In each section it is concluded that the existing literature is deficient and suggestions are made for further research.

There are currently about 3,400 cable television systems in the country. Approximately 85 percent of these systems have fewer than 2,000 subscribers and just more than 1 percent have 20,000 or more subscribers. There are about 10.5 million cable subscribers or about 16 percent of the television homes in the United States. Most of the systems are in smaller markets. The main source of programming at present is network and independent stations.

However, some systems, particularly large ones, originate some of their own programming. These originations are described here because they will be important in the discussion of localism later in this study, because they need to be considered in assessing the effect of regulatory change in large markets, and because more research is needed on the relation of origination to regulatory changes. There are about 500 systems providing local origination to about 7 percent of all television households. Originations are made from systems as small as 180 subscribers to the largest systems and from a few hours a day to very extensive schedules. The programming is a mix of syndicated shows, movies, and local programs. The local programs include city council meetings (27 percent of the originating systems presented these), the "police blotter," local elections, and local sports. In addi-

---

[11] The FCC recently issued new pay-cable regulations but their likely effect is not yet clear. 40 Fed. Reg. 15546-78 (April 7, 1975); FCC Docket of March 28, 1975.

tion, "local access" programs, which are made by people in the community and shown as is, uncensored by the system, are carried on a number of stations.[12]

There is some cooperation among stations in obtaining programs and, although no systematic data are collected on the number of CATV networks, at least two exist. The largest CATV network operates in six midwestern states, supplying programming to 100 cable systems with more than 340,000 subscribers. It provides very specialized programs such as how to cook, how to fix such things as farm equipment and cars. It is supported mainly by sponsors looking for very specific markets.[13]

The effect of unlimited distant-signal importation on cable penetration and profitability has been debated extensively. However, the only study to estimate the effect of additional distant signals on cable penetration is R. E. Park's 1971 study entitled "The Prospects for CATV in the Top 100 Markets." [14] Unfortunately, his estimates were for additional signals under the then existing FCC rules. This creates several problems. First, the FCC constraints on the number of distant signals were binding so that there was little variation in the number of imported signals. As a result, most of the variance occurs in variables not controlled by the FCC, such as price, income, and especially reception. Thus, in Park's study, improved network reception was very significant while the coefficient on imported independents was small and insignificant. Because independent signals will be the only additional signals imported if the current rules are removed, Park's result cannot predict the effect of the change. Second, at the time Park's study was done, the FCC rules required that the nearest signal be imported. Thus, the signals in many markets were not particularly attractive, and so they give a poor indication of the effect of removing all restrictions on distant signals. It is apparent that Park has many unattractive imported independents in his sample, since he obtained the implausible result that an additional educational television station and an additional independent station were valued equally.

Park should have used dummy variables to separate imported signals into several categories according to quality. In this case the FCC rules guarantee substantial variance since some systems are close to large markets with strong independents while others are close to

---

[12] All information from *1975 Local Organization Director* (Washington, D.C.: National Cable Television Association, 1975).

[13] "Monday Memo," *Broadcasting Magazine,* March 22, 1976.

[14] See R. E. Park, "Prospects for Cable in the 100 Largest Television Markets," *Bell Journal of Economics and Management Science,* vol. 3 (Spring 1972), pp. 130-50.

weak UHF stations. This change would give information on the value of strong versus weak signals. The rules also result in some signals being imported only a few miles while others are microwaved over long distances. Adjusting for this would show the value of distant signals relative to close ones of a lesser quality. Not only would these modifications make the study more useful in determining the effect of uncontrolled distant-signal importation but it would be especially valuable in determining the effects of the FCC's recent removal of the antileapfrogging rule.

Although Park's results are not very helpful, qualitative estimates can be made by combining the results of several other studies. R. W. Crandall and L. L. Fray's study "A Reexamination of the Prophecy of Doom for CATV," discussed below, found that the profitability of CATV systems was very sensitive to increased demand, whether reflected in a higher subscription price or in higher penetration.[15] For example, a 2 percent increase in price per year would strongly increase profits. Thus, if additional imported signals increase price or penetration, CATV growth will be stimulated. A study by R. G. Noll, M. J. Peck, and J. J. McGowan found that viewers value additional signals highly, though at a fairly strong diminishing rate.[16] Thus, it is likely that importing additional signals would make cable profitable in some markets where it is not at present and would increase penetration in existing markets.

The cost per subscriber will determine whether distant signals are imported. A recent FCC staff study estimates the cost of importing signals and the increased penetration necessary to make it profitable in different-sized CATV systems.[17] The results show that importation would likely occur in larger systems. For example, for a 10,000-subscriber system penetration would have to increase by 3 percent to 13 percent, depending on distance, to make importation of one station profitable. Since penetration is now fairly low in most larger systems an increase of this size appears reasonable. If the signal is imported over a commercial microwave network, only one to two signals are

---

[15] R. W. Crandall and L. L. Fray, "A Reexamination of the Prophecy of Doom for CATV," *Bell Journal of Economics and Management Science*, vol. 5 (Spring 1974).

[16] Noll, Peck, and McGowan, *Television Regulation*, Appendix A.

[17] An FCC staff paper estimates that microwave transmission is profitable for systems of more than 3,500 subscribers. See "An Economic Evaluation of the Leapfrogging Rules for Independent Stations," mimeographed (Washington, D.C.: Federal Communications Commission, no date). It should, however, be noted that relaxation of the rules would tend to make investment in CATV more attractive and would tend to strengthen the many CATV systems that are now weak financially.

likely to be imported in each market. However, many CATV operators operate their own microwave systems. Currently there are about 350 such private microwave systems averaging about forty to fifty miles in length. In these cases the cost of importing additional signals is low and signals would be imported in bundles, increasing the probability that distant-signal importation would be profitable. In major markets the importation of programming from other major markets would significantly increase the attractiveness of cable. For example, cities such as Philadelphia, which is without a VHF independent, could import a number of signals from New York.

The FCC report finds that growth in smaller markets is not likely to result from relaxed distant-signal importation rules. CATV systems with fewer than 3,500 subscribers at a fixed fee of $6.00 would need increased penetration ranging from 10 percent to 40 percent, depending on distance, to import distant signals profitably. Most small systems already have penetration rates which are quite high so that further increases of this magnitude are not likely. However, these figures are sensitive to increases in the monthly fee. But without reliable information on the elasticity of demand for distant independent signals, no estimate can be made of the increased revenue from higher monthly fees or per-channel fees for the imported station.

The resolution of the copyright issue and the ending of the exclusivity rules are important factors in predicting CATV growth. Naturally, if the exclusivity rules are eliminated, imported signals will be more attractive, cable revenue will rise, and cable will grow. However, this influence will be offset to some extent by the copyright liability newly imposed on the CATV operator. The worst possible case for CATV growth would be the imposition of copyright liability with no relaxation of the exclusivity rules. The result of this case was estimated by B. M. Mitchell.[18] He found that even modest copyright fees would significantly reduce the profitability of CATV, but he was probably too pessimistic. Since such a large part of a CATV system's costs are fixed, small differences in assumptions about cost and demand can make a great difference in profit projections. Crandall and Fray have effectively criticized many of Mitchell's assumptions.[19] They found that Mitchell projected no increase in the subscription fee for fifteen years, overestimated the costs of cable installation significantly, and underestimated ultimate penetration. In addition, Mitchell assumed no future revenues from pay television, but in 1975

---

[18] B. M. Mitchell with R. H. Smiley, "Cable, Cities, and Copyrights," *Bell Journal of Economics and Management Science*, vol. 5 (Spring 1974), p. 235.

[19] Crandall and Fray, "Prophecy of Doom for CATV."

at least sixty systems were operating pay cable. Finally, Mitchell overestimates the economies of scale in cable systems resulting in especially pessimistic predictions for smaller systems. Correcting for all but the pay-cable omission, Crandall and Fray find that cable will do fairly well even with no regulatory change. It does not appear that copyright liability will significantly reduce CATV growth.

If the exclusivity rules are relaxed or removed, cable growth will be stimulated, but we do not know by how much. Park has provided a base to work from by estimating how much of the time the distant signal must be blacked out.[20] He found that the rules could leave distant signals in the top fifty markets blank up to half of the time, but that the rules did not black out more than 15 percent of the programs in the second fifty markets. His estimates are low because he assumes that some substitution will take place, but to date the technical arrangements for this have not proved economical. The amount of blacking out with no substitution should be calculated.

Given that we know the increase in available programming caused by ending the exclusivity rule, we still do not know the marginal value attached to the increase. Reestimation of Park's study of the top 100 markets along the lines suggested would give quantitative results on this. The fact that CATV systems are sensitive to increases in demand does make possible the qualitative prediction that removal of the rules would help the industry.

If the restrictions on cable are relaxed and it starts to grow in mid-size and large markets, what will happen to local origination? There has been virtually no research on this topic. Origination will tend to increase because larger systems can better exploit the economies of scale in programming. This is important because growth would probably promote the FCC policy of localism more effectively than is possible on over-the-air television. There would also be an incentive to reduce local origination because more attractive distant signals would be available. It is probable that there would be a net increase in origination, but more study is needed. To date no estimates have been made of the effect of origination on penetration and of the possibilities for the growth of origination given its cost. The data for this very useful study are available.

In conclusion, more studies are needed to predict the growth of CATV in the event various restrictions are removed. Most useful would be a study similar to Park's study of the "Prospects for CATV

[20] R. E. Park, *The Exclusivity Provisions of the FCC's Cable Television Regulation* (Santa Monica, Calif.: Rand Corporation, 1972); summarized in *TV Communications*, August 1972.

in the 100 Largest Markets" which exploits the information supplied by the variance in the quality of imported signals. The studies need to pay more attention to the cost of importing distant signals than most previous studies have. The relation between market size and CATV system size needs to be established, and the economies of importing distant signals relative to system size and thus to market size have to be investigated in more detail. This work should produce the information necessary to estimate the effect of removing the exclusivity rules. Finally, the relationship between local origination, penetration, and distant signals needs to be studied.

The Economic Effects on Broadcasters

Signal importation by cable makes available more channels for viewing by cable subscribers. This fragments audiences of local broadcasting stations and, hence, reduces their advertising revenues. The reduction in advertising revenues may threaten the quality of local broadcasts or even the viability of some local broadcasting stations.

Perhaps the first question to ask is why adverse effects on local broadcasting stations are relevant. No one holds that the growth of CATV, or any other new technology, should be restricted simply because it endangers net revenues of established companies.[21] However, the FCC had adopted the position that unfettered growth of CATV could be contrary to the public interest in two respects. First, the possible decline in the number of broadcasting stations would be counter to the long-standing FCC policy of localism. Second, financial difficulties for local broadcasters might result in a reduction in their ability to support local public service broadcasting. The importance of these considerations turns on the magnitude of the effects of CATV deregulation on the number and the economic health of local broadcasters.

Unlimited signal importation and elimination of the exclusivity rules would tend to reduce the revenues of local stations in all markets. However, it is not this fact as such but rather its implications for the availability and quality of over-the-air broadcasts that is relevant. The effects on the availability and quality of over-the-air broadcasts in larger markets are not likely to be substantial. This is true for three

---

[21] This point is at least implicitly accepted by broadcasters. For example, in a letter dated October 24, 1975, to F. Lynn May, assistant director of the Domestic Council, Lester W. Lindow, executive director of the Association of Maximum Service Telecasters describes the adverse consequences of cable in terms of a reduction in the quantity and the quality of over-the-air broadcasts.

reasons: (1) the number of channels in these markets is sufficiently large now that the effect of introducing new channels is fairly small; (2) cable penetration is *less* in larger than in smaller markets;[22] and (3) the profit rates of the broadcasting companies are sufficiently great in larger markets to prevent failure as a result of these changes.[23] The revenues of stations in larger markets presumably would decrease somewhat, or grow less rapidly.[24] However, with profit rates on tangible equipment greater than 100 percent per year (as shown in Table 4), cable deregulation would not lead to either a decrease in broadcast quality or a significant decrease in the number of stations.[25]

---

[22] See Noll, Peck, and McGowan, *Television Regulation*, pp. 153-62; and Park, "Prospects for Cable."

[23] Neither original cost nor book value includes the purchase price of the franchise. When the price of the franchise is included, realized rates of return are lower. However, it is figures such as those in Table 4, rather than realized rates, that are relevant to questions about the industry's ability to hold and to attract capital. This is so because the cost of a franchise is entirely a pure rent; franchises would still be available even if their price fell to zero.

[24] Also, UHF independent broadcasting companies may be helped by better reception. See R. E. Park, *Cable Television and UHF Broadcasting* (Santa Monica, Calif.: Rand Corporation, 1971); also appears as "Cable Television, UHF Broadcasting, and FCC Regulatory Policy," *Journal of Law and Economics*, vol. 15 (April 1972), pp. 201-32.

[25] The statement is in rough agreement with a judgment, based on the available evidence, of a group of economists who are knowledgeable about the industry. See Bruce N. Owen, "Memorandum on Deregulation of Cable Television," December 4, 1975. The FCC report "Evaluation of the Leapfrogging Rules," cited above, predicts little effect on small stations. Members of the industry have expressed a contrary opinion. For example, a statement by Paul E. Sonkin, vice president of affiliate research for ABC Television, dated November 6, 1975, contains the following (p. 8): "The foregoing evaluation has *focused on* smaller television markets. At this point in time, CATV growth and its level of penetration has been greatest in these markets. However, the economic principles arising from the advertising practices in the television industry are equally applicable to all television markets. It is simply a question of time and the extent of CATV growth in particular areas before similar effects will be seen in the larger markets as well." The letter from Lester W. Lindow to F. Lynn May, October 24, 1975, refers to "several instances in the 100 largest television markets" where entry by a new broadcasting station was deterred by "existing or reasonably anticipated CATV importation of distant signals." Since, however, it is possible to deal here only with aggregates, the dominant consideration is the extremely high rate of groups of firms. These show that it is unlikely that firms on average will *stop* broadcasting if profit rates fall.

The point to recognize is that any prediction of harm to broadcasting from increased cable activity must be tempered by an understanding that in the larger markets the level of broadcast service (and profits) is higher and the appeal of cable service is lower. Predictions that an increase in cable activities will harm local broadcasters in smaller markets must take into account the fact that it is in precisely these markets that most cable growth has occurred to date (and that, below the top 100 markets, the "protection" afforded under the FCC rules is less significant).

Table 4
PROFITABILITY OF TELEVISION STATIONS

Commercial TV Stations	Number of Stations	Income Before Federal Tax (thousands) (1)	Original Cost of Tangible Property Less Depreciation (thousands) (2)	Percent (1) of (2) (3)
Small market (below 100):				
Network:				
UHF	59	(−$900)	$ 26,584	(a)
VHF	219	33,460	127,954	26.2
Independent:				
UHF	8	(−1,927)	1,847	(a)
VHF	11	(−3,330)	8,503	(a)
Total, small market	297	27,303	164,888	16.6
Large market (top 100):				
Network:				
UHF	58	7,200	45,595	15.8
VHF	244	474,951	379,676	125.1
Independent:				
UHF	52	(−10,345)	42,655	(a)
VHF	20	11,677	46,165	25.3
Total, large market	374	483,483	514,091	94.0
All stations:				
Network:				
UHF	117	6,300	72,179	8.7
VHF	463	508,411	507,630	100.2
Independent:				
UHF	60	(−12,272)	44,502	(a)
VHF	31	8,347	54,668	15.3
Total, all stations	671	510,786	678,979	75.2

a Loss.

Source: FCC data contained in U.S. Congress, House, Subcommittee on Communications of the Committee on Interstate and Foreign Commerce, *Staff Report on Cable Television: Promise Versus Regulatory Performance,* 94th Congress, 2d session, 1976.

Unlimited signal importation might actually help some independent stations in larger markets, especially those few very high quality independent stations that have been said to have "super station" potential. A station whose signals are carried by cable systems into distant markets will market time to advertisers in a way that capitalizes on its larger regional audiences.[26] Also, the station might "sell" its signal to cable systems directly; this will provide an additional source of revenue, at least until regional advertising becomes more widely established. As such stations reach out for a regional market, local advertisers would switch to stations that lack regional aspirations and whose signals were not carried outside their markets. This reassignment of customers would follow the pattern earlier established in network broadcasting whereby the network sells time to national and regional advertisers.

The available evidence does not show which stations will be most strongly affected by unlimited distant-signal importation. The FCC report implies that stations in smaller markets will generally not face competition from strong imported independents.[27] As discussed above, the largest markets will probably not be harmed. The problem is to determine how seriously the middle-sized markets will be affected and how large this "middle size" is.

There have been three major studies that forecast significant harm to smaller local stations from CATV. The first was F. M. Fisher's article in 1966.[28] This article cannot be used to estimate the additional effect of distant signals since this variable was not significant. Generally, the study had the same kinds of econometric problems as Park's.[29] In Fisher's study the constant term explained much of the variation in cable viewership. Measures of the amount of duplication between signals were the only other significant variables. Their coefficients were strongly negative, contrasting to small positive coefficients on the same variables when used to explain over-the-air broadcasting audiences. Also, in projecting the impact on broadcasters the study considered only fractionalization of audiences, not the additional audiences broadcasters get because of cable. Finally, the study's predictions of severe financial impact from CATV growth in

---

[26] The FCC report "Evaluation of the Leapfrogging Rules" and other evidence suggests that advertisers are currently unwilling to pay for exposure to other than local audiences. However, this fact may simply reflect the present absence of regional networks.

[27] FCC, "Evaluation of the Leapfrogging Rules."

[28] Fisher, Ferrall, et al., "Community Antenna Television Systems."

[29] Park, "Prospects for Cable."

small markets were based on 13-year-old data, and they do not seem to have been confirmed by events.

The most extreme forecast of adverse impact from unlimited signal importation appears in a consulting report by Statistical Research Incorporated for the National Association of Broadcasters.[30] The NAB used this report to estimate that over half of the stations in the country would be driven out of business. They based this on an analysis of the audience share of independents in the largest markets. In the largest markets the study projected audience losses to independents of 25 percent in prime time and 61 percent in early fringe. However, for markets with imported independents the study only estimated 33 percent audience loss in early fringe and 16 percent in prime time for local stations. In addition, the one valid regression of CATV growth on local audiences found that there was no significant impact on fringe audiences and that the impact at other times explained only about one-third of the variation in audience size. Thus, the study does not produce soundly based conclusions.

Park's study also forecast significant impact on smaller markets.[31] His forecasts, however, were based on a number of assumptions that probably exaggerated CATV growth and impact. The study was weakened by the assumption that the whole nation could be wired at the same cost per house as those homes already wired. Also, it did not account for competition in the fringes of markets but assumed that all markets were autarkic.

The effect of CATV on UHF television stations is also of importance. The available evidence indicates that cable will help UHF in the short term. The Statistical Research Inc. study found that cable reduced the tuning handicap substantially.[32] Studies by Park[33] and the FCC staff[34] found that in many cases audience fragmentation was more than offset by the reduction of the UHF handicap. The Association of Maximum Service Telecasters claimed in 1971 that the UHF handicap was rapidly decreasing, but the handicap does not appear

---

[30] Statistical Research Inc., "The Potential Impact of CATV on Television Stations," Appendix E to National Association of Broadcasters filing in FCC Docket 18397-A, Fall 1970.

[31] R. E. Park, *Potential Impact of Cable Growth on Television Broadcasting* (Santa Monica, Calif.: Rand Corporation, 1970); summarized in "The Growth of Cable TV and Its Probable Impact on Over-the-Air Broadcasting," *American Economic Review*, May 1971. Also see FCC, "Evaluation of the Leapfrogging Rules."

[32] Statistical Research Inc., "The Potential Impact of CATV," D-1 to D-14.

[33] Park, *Cable Television and UHF Broadcasting*.

[34] FCC, Broadcast Bureau, Research Branch, "The Economics of the TV-CATV Interface," FCC Staff Report, July 15, 1970.

to have been overcome yet. In addition, an examination of UHF stations in Toronto, where there is a great deal of competition on the cable, found that these stations' success was in large part because they were carried on cable.[35] Thus, it appears that UHF stations are not likely to be severely affected by CATV growth, even with distant signals. However, reestimation of the effect of cable on UHF using current data seems necessary to settle the debate.

The greatest impact of cable growth will probably be on VHF stations. Many of these stations are extremely profitable and could absorb a significant impact. The networks, stations owned by the networks, and network-affiliated stations in the top 100 markets are all highly profitable. Some independent VHF stations and most UHF stations might benefit from relaxed rules. The most likely to be hurt are network VHF stations in the middle-sized markets. Thus, an analysis is necessary of profit margins for the stations most likely to be adversely affected. This requires detailed station-by-station projection of operating revenues and costs for the coming five to ten years.

Forecasts of the effects of CATV on local stations are based on relationships which describe penetration and the connection between audience size and advertising. Although it is useful, such information does not go directly to the question of the effect of unlimited signal importation. Examples from particular markets which do address this issue show a mixed pattern. Of the eleven cases cited, three show substantial diversion.[36] However, no stations appear to have closed or drastically cut local service. The same is true of the cases analyzed by Statistical Research Inc. This emphasizes the importance of estimating the effect of distant-signal importation using actual cases rather than a model.

The survival of local stations in the face of cable deregulation would be affected by a number of important additional factors. The networks have the means and an incentive to cushion reduction in local station revenues. As fragmentation of local viewing audiences occurs with the entry of cable, the local affiliate of the networks would experience reduced revenues from local advertising. In response, the networks could increase the revenues from national adver-

---

[35] U.S. Congress, House, Subcommittee on Communications of the Committee on Interstate and Foreign Commerce, *Staff Report on Cable Television: Promise Versus Regulatory Performance*, 94th Congress, 2d session, 1976, p. 47.

[36] Three cases are cited in a letter dated November 7, 1975, from David C. Adams, vice chairman of the National Broadcasting Company, to Paul W. MacAvoy. Five cases appear in FCC, "Evaluation of the Leapfrogging Rules."

tising paid to the local broadcaster and would do so if they their local outlets. Local stations would get an increased share their revenues from the networks and a decreased share from local advertising. The results would not be the loss of local broadcasting but rather a reduction in the profits of the networks. Since the networks and their wholly-owned subsidiaries are not close to failure, the ultimate result is not likely to be a reduction in the number of stations to the extent that this pattern develops.

The existing literature has, apparently without exception, overlooked the question of how broadcasters will respond to competition from cable. Effective response would, of course, limit revenue losses. But, more important than this, competition from cable could tend to make the programming of local stations more truly local. Rather than simply going out of business, broadcasters could respond to cable with strenuous efforts to differentiate and expand their product. Local programming should have some value for audiences, and over-the-air broadcasters may be able to provide this local programming better than cable companies. Thus, in response to cable fragmentation, broadcasters could increase local programming and reduce national or prepackaged programming which is duplicative of that arriving over the cable. The strengthening of local programming in competition with the cable could very well strengthen the financial viability of the local broadcaster.

For all these reasons we have only a very rough idea of the potential impact of CATV expansion on broadcasters. Several new studies are needed. Perhaps the first would update Park's study using the latest work of Crandall and Fray on future CATV growth. This would provide an indication of the effect of CATV systems of about 10,000 subscribers on the broadcasters. If Crandall and Fray's work is extended to analyze the prospects of CATV systems of all sizes, then the effects in all markets could be estimated. Even more useful would be an analysis of the impact of CATV on broadcasters by using actual data rather than a model with separately estimated parts. The Statistical Research study does contain useful data from the Neilsen and Arbitron television rating services. These data give viewing patterns for CATV and non-CATV homes in the same market. Updated and combined with other data, they could provide the basis for a time series and cross-section analysis of the actual effect of CATV distant-signal importation. This would include the adjustments made by local stations, the changes in program costs for purchased programs, and the change in payments from the network.

services would likely experience improved signal
important, an increase in viewing options as a
ng of restrictions on the use of signals. However,
all reduction in the quality and in the number of
over-the air casters, which would reduce the quality of reception of those not on the cable. Some consumers could lose over-the-air reception entirely if they do not subscribe to cable or if cable is not available in their area. The problem then is to assess the trade-off between those who get benefits from greater availability on the cable and those who are hurt by loss of broadcasting services of the cable.

To date only the benefits of another station to cable consumers have been estimated, based on the usual "willingness to pay" methodology. An econometric model by Noll, Peck, and McGowan related the number of CATV subscribers to price, income, the number of households, system age, and a measure of the added diversity afforded by cable.[37] The relationship they derived implies that increasing the number of signals from nine to ten will, for a system with 10,000 subscribers, provide annual benefit of $39,000.[38] Assuming that annual benefits remain unchanged, the present value (at a discount rate of 10 percent) of the benefits stream from providing one additional channel to all cable subscribers in the United States is more than $400 million.

This is only part of the complex "trade-off" problem which needs further research. To date the question of lost service has not been properly treated. First, the actual reduction in the number of stations broadcasting as a result of changes in FCC policy has not been estimated. It is expected that station closings and the greatest reduction in the level of service will occur in the middle-sized markets. But, as indicated below, those are preliminary conclusions that do not describe changes likely to occur in each metropolitan and rural region. The benefits of those who get stations over cable but are now receiving only a poor or limited number of signals have not been indicated in detail as well. Second, changes in FCC policies for over-the-air broadcasting coincident with changes in cable regulation could increase the number of homes receiving over-the-air broadcasts. This could be done by relaxing regulations on the installation of satellite

---

[37] Noll, Peck, and McGowan, *Television Regulation*, p. 297.

[38] This figure is an estimate of the increase in consumer's surplus, including benefits to both current subscribers and new subscribers.

and repeater stations, and by allowing them to originate local services.[39] In addition, over-the-air stations could be allowed to consolidate, thereby creating stronger signals, which would reduce the number of homes unable to receive over-the-air broadcasts.[40] This would provide incentives for stronger over-the-air programming—available through higher purchase prices for better programming, made possible because of economies of scale in program production—thereby maintaining audience by reducing the attractiveness of cable television. A careful examination of these questions could indicate that relaxation of controls on CATV need not reduce the number of homes receiving stations over-the-air.

It is important to note, however, that the changes outlined above could have effects on the nature of television programs. The FCC's policy of localism may be seriously affected. To the extent that localism is reduced, then estimates need to be made of the costs and benefits of such reduced decentralization. First, the nature of the reduction in localism needs to be determined. At present, local news receives good ratings and is often profitable. Apparently, local news originating from repeaters and satellites is possible at low costs.[41] This news source, coupled with the superior coverage of regional stories made possible by financially stronger regional stations, might be preferred by the viewer to present local news. In addition, independent stations offer more local service programming than network stations. The main loss would occur in programs that were not in prime time and were not news programs. Estimates should be made of the economic and social impact of these programs. These losses need to be compared with the benefits obtained by viewers from better reception and from the more specialized and local programs that cable and pay cable are likely to offer.

Another problem is the question of rents and their relationship with supply. Some have noted that many performers receive large rents and so could be expected to supply their services even if the price dropped substantially.[42] These consumers would not experience significant program loss. However, others have noted that com-

---

[39] Paul I. Bortz, Robert C. Spoonberg, Fred P. Zenditte, "Broadband Communications: Rural Areas," Final Report, Prepared for the Office of Telecommunications Policy by Denver Research Institute, March 1973.

[40] H. J. Levin, "Spectrum Allocation without Markets," *American Economic Review*, vol. 60 (May 1970), pp. 209-18 and "Comments," pp. 219-24. Also see Noll, Peck, and McGowan, *Television Regulation*, chapter 3.

[41] House, Subcommittee on Communications of the Committee on Interstate and Foreign Commerce, *Staff Report on Cable Television*.

[42] Noll, Peck, and McGowan, *Television Regulation*.

petition and rents are compatible.[43] What must be established, in order to argue that there will not be significant program-supply effects if prices for entertainment material drop, is rents on the margin. Both R. W. Crandall[44] and B. M. Owen, J. Beebe, and W. G. Manning, Jr.,[45] have analyzed the market for programs. Their approaches are quite different but both conclude that some of the rents to producers and performers are inframarginal. Owen, Beebe, and Manning's approach is potentially productive because it is the most comprehensive and sophisticated analysis so far, but it has not been empirically tested. Of course, all existing programs are fixed in supply and a lower price for them will not affect supply. Rents to networks and to larger stations have been established, and the predictions that reduced demand will only reduce rents, not supply, over some range seems reasonable.

In summary, different groups of consumers would gain and lose from freeing up restrictions on cable television. The gainers would be current subscribers and consumers not now having access to cable who would be willing to subscribe under the less restrictive rules. The losers would be those placing a high value on local programming received over the air and reduced by cable fragmentation. Some consumers may lose service entirely, although this is not likely. Some reduction in the quality of over-the-air broadcast signals may be experienced. The magnitude of most of these effects has to be estimated.

Pay Television

The abundant channel capacity in cable makes it possible to provide programming on a per-channel or per-program pay basis. Unlike over-the-air television, which as a mass media must appeal to mass audiences, the programming on a cable channel can be directed towards more specific, less popular, tastes and desires. Therein lies the crucial advantage of pay cable.

Broadcasters essentially act as brokers who deliver attention of an audience to advertisers. The advertisers' interests lead to programming with a mass appeal, to the exclusion of programs directed to

---

[43] For example, S. M. Besen and B. M. Mitchell. See R. G. Noll, M. J. Peck, and J. J. McGowan, "Economic Aspects of TV Regulation," *Bell Journal of Economics and Management Science*, vol. 5 (Spring 1974), p. 30.

[44] R. W. Crandall, "The Economic Effect of Television-Network Program Ownership," *Journal of Law and Economics*, vol. 14 (October 1971); and R. W. Crandall, "FCC Regulation, Monopsony, and Network Television Program Costs," *Bell Journal of Economics and Management Science*, vol. 3 (1972).

[45] Owen, Beebe, and Manning, *Television Economics*.

specialized interests. At its roots, this situation exists because television does not have a market situation in which viewers can register the intensity of their preference for particular types of programs—it is "watch" or "not watch." Pay cable, in contrast, allows consumers to express the intensity of their preferences by paying for particular programs.

Pay television would benefit consumers as a whole only if it increased the total supply of programs. The supply of some types of programs—for example, the World Series—cannot be increased. Most types of programs, however, are available in increased supply if program prices are increased from the additional demands. It is reasonable to expect that pay cable could lead to an increase in the supply of programs, especially specialized programs.

Unique and popular live broadcast, especially sports events, pose a problem. Their diversion from over-the-air broadcast to pay television would leave the vast majority of consumers without access.

It does not appear so far that movies and other entertainment would be in such restricted supply, however. At present, major distributors are able to schedule pictures first in metropolitan theaters, then in local theaters, then on network television, and finally on syndicated television. Pay-cable exhibition would likely delay the television exhibition of such films. Such films would be released to television after their pay-cable run, in order to capture the advertising revenues from those audiences unwilling to pay to see the movie earlier. This is the present sequence, so that there is no restriction of access likely from cable television as a result of payment for first-run movies.

All these conclusions are tentative. Further research is needed into the programming offered on pay television as it expands, as is further consideration of the Hartford experiment in over-the-air pay television. In that experiment, serious music was one of the most popular offerings, suggesting that the claims of pay-television proponents may be valid.

An extensive theoretical analysis of the trade-off for consumers between gains for cable users from pay television and losses of others has been made by Owen, Beebe, and Manning. They find that under reasonable conditions consumers will be better off with pay television than with free television. However, their conclusion rests on hypothetical assumptions which they have not tested. An empirical test of this model would be required before analysis would support open access for cable to all program materials.

The effect on the poor of pay television is an important question. Noll, Peck, and McGowan have estimated the social value of free television at between $20 and $30 billion a year, and viewing patterns by income group indicated that a proportionately large share of this goes to lower income groups. They note that the higher prices alleged to be the result of television advertising are on goods bought by the poor and so the poor "pay" a significant part of the cost of free television. These are very rough calculations. However, they do suggest that free television is not a "free lunch" even for the poor. They correctly point out that losing free television would have a negative income effect on all over-the-air viewers and especially on poor ones. But these calculations are not helpful to assess the effects of shifting some programs from free to pay television. More detailed work is necessary on gains and losses of viewer groups before these effects can be evaluated.

Educational Television

The effect on educational television and public broadcasting of changes in cable regulation needs to be examined. Educational television (ETV) stations suffer all the handicaps of UHF signals and usually have lower broadcast power as well. The research on the UHF handicap suggests that ETV would be helped by the expansion of cable by gaining much better viewer access; but whether it would be helped more than it would be hurt by the importation of distant signals is not known. In addition, ETV and PBS revenue is not closely related to audience size, so the effects of fragmentation on station viability are difficult to assess and they have not been assessed. Moreover, the FCC now stops the importation of distant ETV signals if a local station or educational authority can show harm; the unlimited importation of distant signals would help those ETV stations now foreclosed by FCC rulings, but it is not known "which" would be helped and "how much" of an economic difference it would make.

An important consideration here is the level of marginal costs of importing additional signals. If marginal costs are "low," then importation of ETV signals could be economical. Some ETV stations would have bigger signals, some ETV stations would have bigger audiences, and some homes would receive ETV for the first time. The larger audiences for some stations might increase donations and would permit exploitation of economies of scale in audience size. Thus, the costs and benefits to ETV of more signal importation are highly problematical at this time.

3

ECONOMIC POLICY RESEARCH ON CABLE TELEVISION: ASSESSING THE COSTS AND BENEFITS OF CABLE DEREGULATION

S. M. Besen, B. M. Mitchell, R. G. Noll,
B. M. Owen, R. E. Park, J. N. Rosse

Introduction

On April 2, 1976, Paul W. MacAvoy issued a memorandum to the Domestic Council Review Group on Regulatory Reform titled "Status Report on Proposals for Regulatory Reform in Broadcasting."[1] Particular attention was paid to the desirability of reducing or eliminating anticompetitive regulations inhibiting the development of cable television. The document concluded that the knowledge generated by existing research was insufficient to bear the burden of proof for a decision to deregulate. Comments on that research and suggestions for further study were included in the memorandum.

In the interest of furthering public discussion of the issues raised, the Office of Telecommunications Policy engaged the six authors of this document to undertake the following tasks:

1. Review MacAvoy's assessment of the state of knowledge.

2. Develop an independent description of deficiencies in data and analysis.

3. Provide a candid assessment of the utility and the necessity of further study.

4. Propose, if necessary, a research program to obtain the information essential to regulatory reform.

All six of us have been associated with the body of research referenced in the MacAvoy memorandum and five of us have made

---

This paper was prepared for the Office of Telecommunications Policy, October 1976.
[1] The memorandum is presented as Chapter 2 in this volume. Hereinafter it is referred to as "MacAvoy."

extensive contributions to it. This document is our modest response to the charge put by OTP, a modesty that is caused partly, we confess, by a certain sense of futility.

The Burden of Proof. This sense of futility is stimulated by the apparent burden of proof placed on policy research in the broadcast area. The process of public policy making quite rightly asks who will be the gainers and the losers from a proposed change and how do gains compare to losses. It is in the nature of policy making to seek the most detailed possible answers to these questions. MacAvoy's central complaint is that the detail afforded by the existing research is insufficient.[2] We agree that the level of detailed knowledge is neither as great nor as reliable as one would like, that existing research could and probably should be refined to provide better answers, and that new research could provide marginal improvement.

The process of comparing gains and losses is generally accepted as proper for such governmental matters as changes in the tax code, provision of public services, and welfare. We as a body politic are accustomed to decisions being made on that basis even when the losers represent substantial segments of the population or important forms of economic activity.

Such a burden of proof is not ordinarily put on innovation in the unregulated industrial and commercial sectors of our economy, however. No public agency required the producers of nylon or transistors to establish that the social benefits of such products outweighed costs, in spite of obvious impacts on consumers and on a wide range of preexisting industries. To have done so would have been an inappropriate constraint on the vitality of our enterprise economy. Given the difficulty of establishing the needed case in advance, the burden-of-proof requirement would almost certainly have led to the suppression of the innovations.

Even in the case of regulated industries, the burden of proof has not always been put on proponents of innovation. So far as we know, proponents of television were not compelled to establish that social benefits exceeded costs in the early postwar period.[3]

---

[2] MacAvoy, for example, wants calculations regarding the effects of cable deregulation on specific television stations in specific markets.

[3] In that case, of course, potential earnings from television were much greater than from radio, and many of the commercial interests benefiting from television were coincident with potential losers in radio. In one other situation television was not required to bear the burden of proof. The demise of minor league baseball was in large part the result of the carriage of the telecasts of major league baseball games into minor league cities. The effect of these telecasts was to "fragment" the (live) audiences at minor league games, to reduce team

One is entitled to ask, therefore, why such a burden of proof should be imposed in the case of current broadcast policy where, again, a newer technology (cable) may have significant impacts on an older one (over-the-air television). Three possible answers suggest themselves.

First, the fact that over-the-air broadcasting is a regulated industry means that its members have access to administrative and judicial processes which can be used to impede or delay change and which must, therefore, be dealt with before accommodation to technological change can be made. In other words, the interests created by prior regulatory decisions must either be compensated or be convinced that further delay is futile.[4]

Second, it is alleged that potential economic profits created by existing regulatory practice are used by regulators to subsidize provision of goods that might not otherwise be produced.[5] Local origination, the range of phenomena encompassed by the policy of localism, and public service broadcasting appear to fall in this category. In spite of cable's own capacity for public service broadcasting and local origination, the unregulated expansion of cable television may, by reducing stations' profits, reduce the provision of such services by broadcasters.

A third possible answer has to do with potential effects of income redistribution. The expansion of cable television could reduce the viability of over-the-air broadcasting to the point where some stations would leave the air and/or the quality of over-the-air broadcasts would decline. In either case, nonsubscribers in affected markets would be less well-off than in the absence of cable, and an implicit income redistribution would have taken place from nonsubscribers to subscribers and, if the nonsubscribers happened to be low-income households, a burden of proof to demonstrate that the net effects were socially desirable might be created.

revenues, and eventually to cause some teams to cease operations. Minor league teams were not compensated for the losses they experienced, nor were the interests of those fans who preferred live minor league baseball to televised major league baseball taken into account.

[4] For discussion of this problem, see Richard B. Stewart, "The Reformation of American Administrative Law," *Harvard Law Review*, vol. 88 (1975), p. 1667; Louis L. Jaffe, *Judicial Control of Administrative Action* (Boston: Little, Brown and Co., 1975); Richard A. Posner, "Theories of Administrative Regulation," *Bell Journal of Economics and Management Science*, vol. 5 (1974), p. 335.

[5] See W. S. Comanor and B. M. Mitchell, "The Costs of Planning: The FCC and Cable Television," *Journal of Law and Economics*, 1972; see generally Richard A. Posner, "Taxation by Regulation," *Bell Journal of Economics and Management Science*, vol. 2 (Spring 1971).

We will take the above as justifying the asserted burden of proof, without necessarily accepting its desirability. We will address our comments to the relevance, availability, and reliability of existing research and to the potential usefulness of additional research.

Cable Is Different from Other Regulated Industries. Unfortunately, the nature of the problem of broadcast policy makes impossible the acquisition of information having the detail and the reliability of, for instance, information about motor freight transport. In trucking, researchers deal with a well-known, stable technology, which is currently unthreatened by new technologies and about which considerable public information exists. Trucking services are bought and sold, so that price and quantity information is readily available. Trucking services are homogeneous products unaffected by the technology of delivery; as a result, the policy researcher need not be concerned with whether deregulation will result in trucks using turbines instead of piston engines. Regulatory objectives in trucking are clear-cut so that an evaluation of the effectiveness of regulation (or its absence) is straightforward. Finally, since trucking regulation essentially involves certification of routes and rates, the researcher has a limited set of policy alternatives to evaluate.

In contrast, the current policy issues in broadcasting stem from the rapid evolution of new technologies. Little public information on the economic aspects of these technologies is available and, even for traditional methods of broadcasting, the available information is fundamentally deficient. The broadcast industry's product for viewers is not priced; as a result, researchers can measure its value in use only indirectly, if at all. Moreover, the fundamental building blocks of industry analysis—information on costs and profits of firms—are available only in such unreliable form as to make their detailed analysis fruitless. By contrast with trucking, in the broadcast industry the nature of the delivery mechanism itself has an important effect in determining the characteristics of the materials (programming products) that are conveyed. Finally, the very range of policy alternatives that a researcher is called upon to evaluate seems limited only by the imagination of the adversaries.[6] The objectives of broadcast regulation have not been stated in a sufficiently succinct way to permit

---

[6] A related point is that it is by no means clear, in the case of cable television, what *deregulation* means. Does it include the FCC's distant-signal importation restrictions, including the exclusivity and leapfrogging rules? Does it include some or all of the restrictions on pay television? Does it include the FCC certification of cable systems? Of technical standards? The list goes on.

research that can evaluate quantitatively the effectiveness of regulatory practices.

In short, the task of supplying reasonably accurate information about how many common-carrier truckers will at what prices be serving Grand Rapids, Michigan,[7] as a result of deregulation is easy when compared to the task of predicting how many over-the-air and cable broadcast facilities will be located in Grand Rapids, supplying at prices to be predicted what character of programming to what fractions of the population from what local origination, network, syndicated, or distant-signal sources as a result of a yet-to-be specified subset of deregulation proposals, using a technology that may well evolve in new directions as a consequence of whatever policy change is made.

A Roadmap for Analysis of Deregulation. In order to estimate the impact of unregulated cable television on the welfare of viewers and broadcasters in specific markets, the following individual pieces of information would be needed:

1. The number of subscribers to cable systems;

2. The services offered by cable and the diversion of audience from local stations to imported stations or other sources of programming on the cable system; and

3. The effect of audience diversion on the revenues and profits of existing stations.

MacAvoy proposes case studies for providing insight into the impact of cable, but case studies are probably not an appropriate methodology.[8] The best method for acquiring the needed information is to collect and use the data generated by existing cable systems and by broadcast stations. Those data could be used to estimate statistical demand and cost functions for both types of enterprise, at least for

---

[7] The example could just as easily be airlines in Plains, Georgia, or natural gas suppliers in New Haven, Connecticut.

[8] Case studies of station performance provide an alternative to statistical analysis. However, the experience in each case is the result of many causal factors. For instance, cable impact, incompetent management, accounting peculiarities, or location in an exceptionally poor media market are all possible causes for the unprofitability of a given case-study station, but only the first is relevant to cable policy. Causal factors can only be identified and quantified reliably by statistical methods designed to quantify the average experiences of a large number of cases. Further, case studies do not permit the construction of the policy experiments needed for policy evaluation. Case studies may sometimes be a useful *supplement* to statistical analysis, but they cannot *substitute* for such analysis.

the range of regulatory alternatives encompassed by historical experience. Given the differential regulatory treatment of markets, as well as past changes of policy, a range of policy alternatives could be analyzed. The range could be increased by future experiments in test markets.

The first step in a statistical analysis of the impact of cable deregulation would be to estimate the demand for cable, as determined by the service offered by the cable system, the amount and quality of service available from over-the-air broadcasters, the price of the cable service, and other relevant features of the local market, such as personal income and entertainment options apart from broadcasting. The result of this analysis would be a prediction of the penetration rate of cable systems in each service area, given assumptions about the prices and services offered by the cable operators.

The second step in the analysis would be to determine the extent to which cable systems are viable economic enterprises, given the demand relation that has been estimated. A necessary ingredient of this analysis is an estimate of the costs of providing cable service, as determined by the physical characteristics of the neighborhood to be wired, the number of homes that will choose to subscribe, and the programming services to be offered (including copyright liability for their use).

These demand and cost relationships can then be used to predict the number of houses that will subscribe to cable systems in each market. Each neighborhood can be characterized in terms of the quality and the number of off-the-air signals its residents can receive, the costs associated with constructing a cable plant, and the other features of the neighborhood that will affect its demand for cable. For each neighborhood, one can then answer two questions: first, would any configuration of price and service make cable profitable in the area; and second, if so, what configuration would make cable most profitable or would maximize the number of subscribers, given that cable profits are held to a fair-rate-of-return limit?[9] By totaling the results over all neighborhoods, the analysis will then provide an estimate of the fraction of a market that will be offered service and, within that fraction, the range of penetration—that is, the proportion that will subscribe from among those offered service—that will occur.

The third step in the analysis would be to examine the viewing patterns of cable subscribers to determine the extent to which broad-

---

[9] Rate-of-return regulation of cable by state and local authorities, whether desirable or not, seems to be a realistic assumption for any such analysis.

casters lose audiences because of the existence of more viewing options on cable systems. Stations with strong signals and large audiences are likely to suffer audience diversion, and their revenues are likely to decline. For stations with weak signals, resulting from UHF problems or the nature of the physical terrain, audience diversion because of competition may be more than offset by improved signal quality on the cable. Two factors are relevant to this analysis. The first is the distribution of viewing by cable subscribers among the services offered on the cable system, as determined by the number, type, and quality of these services. The second is the extent to which existing cable audiences differ in their preferences from those who would be subscribers to a more extensive cable service. Because cable operators rationally offer service first to groups most interested in receiving cable services, this may generate distributions of the cable audience among types of services which are not typical of what can be expected under deregulation.

The fourth step would be to convert the estimates of audience diversion to loss of revenues incurred by stations. This analysis has three elements.

The first element is the relationship between the local audience of a station and its revenues, which determines the net change in revenues because of more competition coupled with a clearer delivery of its signal on cable in the home market. Another element is the relationship between the audience a station captures on a distant cable system and its revenues. More competition from cable systems in the home market will, for some stations, be at least partly offset by greater audiences in distant markets.

The third element is the relationship between competition and programming, and its effect on revenues. Stations may respond to greater competition by changing their programming format, altering both their costs and their over-the-air audience. Some, such as strong VHF independents, may see in distant cable audiences the opportunity to become more like a network and, therefore, may upgrade their programming; others, in the face of declining revenues, may cut costs by offering lower quality programs. Still others may change their target audience, say, from devotees of old movies to particular ethnic groups.

A detailed analysis of the revenues of stations, based upon both the number and the types of viewers of the station's programs, would, when combined with the data about audience diversion, provide an estimate of the revenues of stations in a postcable environment.

The penultimate step in the analysis of the impact of cable would be to determine the relationship of revenues to profits, either indirectly by first estimating station cost functions, including the costs of alternative programming formats, or directly by observing for each type of station the relationship between revenues and the prices at which stations change hands. Among the elements of this analysis would be an assessment of the true costs of each category of programming, stripped of whatever economic rents current cost estimates may contain. This analysis would then enable predictions to be made of the stations that would no longer be economically viable when facing cable competition of the kind likely to emerge in their home markets, and of the stations that would remain viable only by eliminating local programming, weakening their program offerings, or moving towards regional or national network status.

The final step in the analysis would be an assessment of the effect of all these changes on consumer welfare. Households not subscribing to cable could be affected by cable deregulation if the stations they view were either to go off the air or to switch their formats. Households subscribing to cable would be affected by the greater range of service offered by cable, the change in formats of local stations, and the expense of accepting service. An analysis of the demand for cable can shed light on the magnitude of these effects through its estimates of the effects of alternative price and service combinations on the willingness to subscribe; however, without market transactions to observe, even the rough magnitude of the effect on nonsubscribers can never be more than conjectural.

What Can Further Analysis Accomplish? The remaining sections of this report discuss the state of existing and potential knowledge on each of these major components of an impact analysis of each cable television. Each section summarizes the information contained in the existing literature, the additional information that might be acquired from further research, and the information that is probably impossible to obtain for the particular point of analysis under examination. Before proceeding to these more detailed discussions, two especially important observations should be made.

First, the utility of further analysis on one issue is limited by the state of knowledge on others. A station-by-station impact analysis of unregulated cable service can be only as good as the quality of the weakest component of that analysis. Thus, from a policy—though not necessarily from a scientific—perspective, it makes no sense to refine research knowledge on one element of the analysis if another element is woefully weak and impossible to improve through additional research. At present, the weakest links in the chain are information on the costs and profitability of broadcasters and on the demand and costs of new programming services that either cable systems or over-the-air broadcasters might offer if cable were deregulated. Neither topic is able to be researched, either because no data exist or because the data are not available to the research community. The FCC data on the costs and profits of broadcasters are next to meaningless and are probably beyond redemption. Cable has not developed sufficiently to cause any significant changes in program offerings by either cable systems or broadcasters, with the single and recent exception of pay-television channels on some large cable systems.

Second, the existing literature is sufficient to answer a more limited but nonetheless significant range of questions. The demand for cable, the costs of cable, the diversion of audiences because of cable, and the effect of audience size on station revenues are well enough known to provide a market-by-market prediction of the likely growth of cable and the effect on broadcast revenues from cable deregulation, assuming that cable systems and broadcasters continue to offer the types and the quality of services that both have offered in the past.

A complete market-by-market analysis has not been done, but it could be carried out by staff members of the relevant government agencies using the tools available. In fact, several existing publications work out this exact problem for a particular market or for hypothetical illustrative examples. To accomplish the task of market analysis, all that is needed is to assemble basic data for each market. Of course, the task is elaborate and time-consuming. The estimate of cable penetration must be built up from an examination of the viability of cable on a neighborhood-by-neighborhood basis, reflecting intramarket variations in income, signal quality, and the cost of constructing cable systems.

Although a complete, nationwide market-by-market analysis has not been done, the general outcome of this mountainous task is reasonably clear, and it is outlined below. One piece of evidence bearing on the question of cable's impact is the experience of broadcasters in television markets outside of the 100 largest markets. Prior to 1972, the FCC did not regulate cable television in these markets, yet between 1962 and 1972, the period in which cable television grew to prominence, the number of television stations on the air in these small markets did not decline.[10] In addition, the evidence on the effects of audience diversion presented in detail below indicates that few stations in the top 100 markets will be forced off the air by cable competition.

An important point to realize is that the audience-diversion argument really works against those who would block cable development. Audience diversion can be important enough to threaten many broadcasters only if nearly everyone in a market subscribes to cable *and* shifts his viewing to new cable services. Yet, if this phenomenon occurs, it is because cable services are generally regarded as superior by most households—in which case there is little reason to protect over-the-air broadcasters.

Conclusions. The conclusions that can legitimately be drawn from the discussion in this report and from the current literature on cable and broadcast economics are:

1. Cable television does not offer an immediate threat to the general public: it will not in the foreseeable future drive over-the-air broadcasting out of business. Instead, it does offer the prospect of giving its subscribers substantially more entertainment options than they now possess.

2. Cable television does pose an economic threat to broadcasters, but only because it threatens a monopolistic position. More competition will result in lower profits for the television industry, and consequently the value of television licenses will fall. But these effects will not be great enough to cause a wholesale reduction in the number of broadcasting stations.

3. The existing literature provides sufficient information to enable government officials to undertake a market-by-market impact analysis of cable television, provided that the objective is limited to estimates of the effect of cable on television station revenues and that the issue of interest is the impact of cable under the assumption that neither cable nor broadcasting change the nature of their services as cable grows. However, detailed predictions of which stations will fail must await the development of reliable cost and profit data for broadcast stations.

---

[10] The number of commercial stations outside the top 100 markets grew from 275 to 277 between 1962 and 1972, while the number in the top 100 markets grew from 302 to 424. Cable penetration outside the top 100 markets averaged 32 percent of all households in those markets in 1973. It is possible that the number of television stations outside the top 100 markets would have grown substantially without cable, but the important point is that existing stations were not driven off the air in significant numbers.

4. Scientifically interesting research questions remain unanswered on issues pertinent to the policy debate over the future of cable television. However, successful prosecution of these research projects is not likely to change materially the overall conclusions that can be derived from existing literature.

5. Among the potentially fruitful research issues that have yet to be thoroughly explored, and which we recommend should be undertaken, are:

- Reestimate a cable-demand model, improving the specification of relevant variables and their functional relationship, and using a new and larger individual household data base, in order to provide more reliable estimates of the quantitative effects of distant signals, local origination, and pay television.

- Using newly available data from the Arbitron rating service, study viewing patterns disaggregated by type of station and by type of viewer for both cable and noncable households, so that the effect of cable viewing options and the resulting impact on local stations' audiences can be measured in greater detail than in the past.

- Use FCC data on television station sales prices to estimate the effect, if any, on profits or expected profits of cable growth.

- Undertake a comparative economic analysis of the history of over-the-air broadcasting in a sample of markets with differing histories of cable development and instances of station failures, the object being to determine the effect, if any, of high-quality cable service on the observed frequency of station failures.

The Demand for Cable

The economic analysis of cable television policy logically begins with establishment of the determinants of the demand for cable television service, commonly measured by the penetration rate: the number of households that subscribe to cable service as a fraction of those offered service. Five major econometric studies of the demand for cable television have been undertaken, about one each year, beginning with the study by W. S. Comanor and B. M. Mitchell.[11]

---

[11] The citations of the five studies are found in Tables 5–9. All subsequent citations of Comanor and Mitchell in this chapter refer to their *Bell Journal* article which is outlined in Table 9.

Theoretically, a large number of factors should affect cable demand—including especially the service alternatives (numbers of signals of each type of station) both on cable and off-the-air, the reception quality of local signals, subscription price, and demographic characteristics of households. However, reliable determination of the quantitative importance of these factors is limited by the accuracy with which some variables are measured and the lack of significant variation across systems in others. The principal findings of each study are gathered together in Tables 5 to 9 in a format that permits direct comparison.

Successive econometric studies have been made with increasingly reliable and detailed data; in addition, more sophisticated statistical and modeling methodologies have been employed. With the single exception of the Hopkins Project study, all penetration estimates are based on data generated prior to the imposition of exclusivity and leapfrogging restrictions on imported distant signals. To the extent that these rules reduce the attractiveness of cable to potential subscribers, the estimates of cable penetration summarized below are optimistic.

When we analyze these studies in light of the data and the methods used, substantial areas of agreement emerge:

First, the reception quality of off-the-air signals is a key determinant of cable penetration. The Hopkins Project study measures quality directly, by field tests, and gets very strong results. Good proxy measures of quality also pick up strongly significant effects (note the Park study).

Second, demand is decidedly price-sensitive; the most recent estimates are of roughly unitary elasticity in the observed range of subscription rates. This value may contain some upward bias because of the simultaneous determination of regulated fees by expected penetration, as suggested by R. W. Crandall and L. L. Fray,[12] but attempts to correct for the bias have been unsatisfactory.

Third, cable demand matures rapidly to a steady-state level within a year or so after service is offered to customers in a specific area. The earlier findings of slower growth are probably because of observing aggregate results of systems built in stages; large systems are built in segments over time so that system-wide penetration reaches a mature level only after several years. Because financing and construction costs can be phased to the opening of new segments, rapid

---

[12] R. W. Crandall and L. L. Fray, "A Reexamination of the Prophecy of Doom for Cable Television," *Bell Journal of Economics and Management Science*, vol. 5 (Spring 1974).

Table 5

HOPKINS CABLE PROJECT STUDY

Author and reference	Hopkins Cable Project—K. Lyall, R. Duncan, and C. F. DeKay, "Estimation of an Urban Cable Demand Model and Its Implications for Regulation for Major Markets," March 1976, The Center for Metropolitan Planning and Research, The Johns Hopkins University, Baltimore, Maryland.
Sample	175 observations for 1974 (distinct "management areas" or construction segments) of 20 cable systems located in 18 southern, midwestern, and western cities; 17 of the systems are in top 100 markets; median penetration is 30%; all have 3-network service off-air. Data gathered by interviewing system operators.

Elasticities
  Price   −1.33
  Income   0.54

Programming
  Duplicate networks   −, insignificant

  Independents
    Class A (large NWC) −, significant
    Class B (small NWC) +, insignificant

  Local origination   +, significant; Δ penetration = .016/channel

Signal quality
  Measured quality   −, significant

  Distance proxy   —

  UHF   +, significant

Other variables   educational, foreign language stations
                        marketing expenditure
                        topography
                        persons per household
                        vacancy rate
                        color set penetration

Methodology
  Dependent variable   log(Pen/1-Pen), i.e., logit

  Equation form   log-linear in independent variables; service variables nonlinear in quality parameters

Table 5 (continued)

Price variable	1967 dollars; monthly service + pro-rated installation fee

Service variables

$$S = \frac{1 + \#\text{cable stations}}{1 + \sum_{1}^{nv} D_{vi} + \sum_{1}^{nu} D_{ui}}$$

for networks, duplicate nets, independents, educational, foreign language

Quality variables

$$D_v = 1 - (1 - SS_v^n)^{1/n}$$

$$D_u = \epsilon \cdot UHF \cdot [1 - (1 - SS_u^n)]^{1/n}$$

SS = signal strength measure, in decibels of attenuation

n = estimated signal quality parameter

ϵ = estimated UHF handicap parameter

UHF = % of households with UHF tuners

Author's conclusions

1. Penetration very sensitive to improvements in network signal quality by cable.
2. Addition of other (independent) signals not important in their urban sample.
3. Demand is at least unit price-elastic.
4. With effective marketing, stable penetration is achieved rapidly, in some 90 days.
5. Substantial local origination increases penetration.
6. Within sample standard errors for penetration are ±6 percentage points; necessarily larger forecast errors are probably comparable to Park's.
7. Current FCC regulations only marginally affect cable expansion.
8. Local regulation (especially rate-of-return and requirements for 100% coverage) severely impede cable growth.

Forecasts

For Baltimore and Cleveland, penetration 23–25%. Park's equation gives 13–15%. Differences attributed to (1) measured signal quality worse than given by distance proxy, (2) adjustment for vacancies, (3) substantial local origination assumed.

Table 6

CHARLES RIVER ASSOCIATES STUDY

Author and reference	Charles River Associates, "Analysis of the Demand for Cable Television," CRA Report #178-2, April 1973, Cambridge, Massachusetts.
Sample	735 individual household interviews (no dates) for 13 systems in 12 states, all but one in top 100 markets; mean penetration 39%. Data gathered from marketing surveys for individual systems.

Elasticity
    Price   −1.09, robust to specification changes
    Income   +, significant

Programming
    Duplicate nets   +, significant; ΔP = .04/duplicate network
    Independents   +, 10% level of significance; ΔPen = .02/independent

    Local origination   +, insignificant

Signal quality
    Measured quality   Off-air signals of less than "good" quality not counted in number of off-air service alternatives. Signal quality self-assessed by household, method not explained.

    Distance proxy   —
    UHF   No distinction made where off-air signal is "good."

Other variables   Length of residence; family size; occupation; age; income distribution
                            Number of television sets; UHF tuner; color; type of antenna system age; marketing expense per household

Methodology
    Dependent variable   H = 1 if subscriber, 0 if not
    Equation form   Linear probability model; estimated by OLS
    Price variable   Monthly fee for basic service
    Service variables $\text{Log } (1 + N_{cable}/1 + N_{off\text{-}air})$
    Quality variables   —

Author's conclusions
1. Unitary price elasticity of demand.
2. Duplicate networks have largest effect of increasing penetration.
3. Independents have half the effect of duplicate nets.
4. Local, nonautomated origination has statistically unreliable effect, but is roughly comparable to an additional independent.
5. Maturation is achieved rapidly, in a year or less.

Table 7

NOLL-PECK-McGOWAN STUDY

Authors and reference	R. G. Noll, M. J. Peck, and J. J. McGowan, *Economic Aspects of Television Regulation* (Washington, D.C.: The Brookings Institution, 1973).
	Model I
Sample	31 cable systems for 1969 with at least 10,000 subscribers
Elasticities	
Price	−.75 to −.9 (calculated in Besen and Mitchell review)
Income	+.75 to +.9 (calculated in Besen and Mitchell review)
Programming	
Duplicate nets	+, significant
Independents	+, significant
Local origination	+, insignificant
Signal quality	
Measured quality	—
Distant proxy	—
UHF	—
Other variables	Educational station dummy variable; all observations deflated by square root of system age
Methodology	
Dependent variable	Log (1 − price/income)/log pen
Equation form	Log-linear
Price variable	Annual fee + 1/5 installation fee
Service variables	$S = \dfrac{1 + \# \text{ cable}}{1 + \# \text{ off-air}}$
Quality variable	—
Authors' conclusions	1. Network programming is highly valued as compared to programming of independents.
	2. Welfare increases, but at rapidly diminishing rate with addition of either networks or independents. Average household would pay 5% of income rather than forego 3-network service.

Table 7 (continued)

	Model II
Sample	35 systems for 1971 submitting completed questionnaires mailed to 400 systems in 3-network markets.
Elasticities	
Price	−1.93
Income	1
Programming	
Duplicate nets	
Independents	$\Big\{$ +, 15% level of significance
Local origination	
Signal quality	
Measured quality	—
Distance proxy	—
UHF	—
Other variables	Top 100 market dummy variable system age
Methodology	
Dependent variable	Pen
Equation form	Linear
Price variable	Annual fee + 1/5 installation fee
Service variables	Single variable for all signals
	$S = {}_{\log} \left(\dfrac{1 + \#\ \text{cable}}{1 + \#\ \text{``good'' off-air}} \right)$
Quality variable	—
Authors' conclusions	1. Profit-maximizing price not significantly different from observed mean of about $5.00 per month.
	2. Rapid maturation—roughly two-thirds of ultimate penetration achieved in 18 months.

attainment of the equilibrium penetration rate is the correct basis for modeling the financial effects of cable operations.

Fourth, the most recent detailed studies of cable demand in urban portions of the top 100 markets estimate that equilibrium penetration will be substantially lower (reaching only 20 percent to 40 percent) than has been observed in small markets and in systems constructed in areas with poor reception. Using quite different data

Table 8
PARK STUDY

Author and reference	R. E. Park, "Prospects for Cable in the 100 Largest Television Markets," *Bell Journal of Economics and Management Science,* Spring 1972.
Sample	63 systems (July 1971) with at least 3 good-quality off-air signals; data verified with system operator.
Elasticities	
Price	−1.01, significant
Income	0.96, significant
Programming	
Duplicate nets	+, significant
Independents	+, significant at 10% level
Local origination	—
Signal quality	
Measured quality	—
Distance proxy	Nonlinear, significant
UHF	Significant
Other variables	System age
	Color sets
	Number of primary network and Canadian signals
Methodology	
Dependent variable	Log [Pen/(Exp(−a/Aga) − Pen)]
Equation form	Log-linear
Price variable	Monthly charge plus annualized installation fee
Service variable	$$S = \frac{1 + \#\ \text{cable signals}}{1 + \sum_{1}^{nv} (1-d^{\delta})^{1/\delta} + \gamma \text{UHF} \sum_{1}^{nu} (1-d^{\delta})^{1/\delta}}$$ Where d = distance from transmitter UHF = % of UHF households γ = estimated UHF handicap factor δ = estimated signal quality parameter
Author's conclusions	1. Reception problems, particularly for UHF, increase penetration. 2. At edges of major markets, penetration varies considerably, according to signal quality. 3. Distant signals increase penetration 5 to 10 percentage points. 4. Income and color set growth will increase penetration over time. 5. Systems mature within 18 months.

Forecasts	With 3 distant independents:
	  St. Louis 26–28% (middle-edge)
	  Salt Lake City 25–29%
	  Evansville 29–63%

Table 9
COMANOR-MITCHELL STUDY

Authors and reference	W. S. Comanor and B. M. Mitchell, "Cable Television and the Impact of Regulation," *Bell Journal of Economics and Management Science,* Spring 1971.
Sample	149 systems drawn from statistics reported in 1970 *Television Factbook;* mean penetration is 63%.

Elasticities

Price	−.1 to −.3, insignificant
Income	+.15, insignificant

Programming

Duplicate nets	−, insignificant
Independents	+, insignificant
Local origination	+, insignificant (nonautomated origination)

Signal quality

Measured quality	—
Distance proxy	—
UHF	Alternative dependent variables

Other variables	System age
	Top 100 market dummy
	Primary network and educational stations

Methodology

Dependent variable	Log (pen)
Equation form	Log-linear
Price variable	Monthly charge for basic service
Service variable	Log $(1 + N_{cable}/1 + N_{off\text{-}air})$, where $N_{off\text{-}air}$ is alternatively defined for different VHF and VHF+UHF A and B contour signals
Quality variable	—

Authors' conclusions	1. For typical systems penetration of 53% in top 100 markets predicted with 5 independent signals.
	2. 80% maturation achieved in 4 years.

bases, these studies (Park and Hopkins Project) reach similar levels of predicted penetration: between 15 percent and 25 percent for the Baltimore and Cleveland markets.

In contrast to the consensus of research on the points just enumerated, reliable estimates of the importance of other determinants of demand are not available.

Fifth, the effect of additional independent signals on cable demand is found to be positive in several studies,[13] but the magnitude of the effect is, as MacAvoy points out, in considerable doubt. Although cable operators, by their marketing decisions, attribute considerable importance to being able to offer distant signals, the most recent econometric studies find only modest effects of imported independent stations on cable penetration. The one limited attempt to control for the program quality of imported independents (Hopkins Project) gave perverse results which may stem from the fact that the sample was limited to only twenty systems, and that, unlike earlier studies, during the period analyzed imported signals were subject to leapfrogging and exclusivity restrictions. Among the other studies, the additional cable penetration that would be achieved by importing four independent signals ranges from 9 percent (Charles River Associates) to 43 percent (Noll-Peck-McGowan) of the households offered service. Because of this high variability, the expected effect on broadcasters of relaxing distant-signal carriage rules should be calculated for several penetration assumptions.

Sixth, local origination may increase penetration to a small extent. However, except for the Hopkins Project study, no statistically significant effect has been measured.[14] Furthermore, cable operators provide relatively little origination except where required to do so by the franchise authority, suggesting that the returns in increased penetration are outweighed by added costs.

Seventh, the extent to which pay-cable channels increase cable penetration has not yet been investigated. Furthermore, in contrast to the demand for basic cable service, the factors that determine the demand for pay-cable programming have not received serious study. At this point there are about 800,000 pay-cable subscribers. Data on

---

[13] MacAvoy cited only the Park study. Others include Comanor-Mitchell and Noll-Peck-McGowan.

[14] Several other studies, including Comanor-Mitchell and Noll-Peck-McGowan, have included an origination variable and found it insignificant. See also R. E. Park, *Potential Impact of Cable Growth on Television Broadcasting* (Santa Monica, Calif.: Rand Corporation, 1970); summarized in "The Growth of Cable TV and Its Probable Impact on Over-the-Air Broadcasting," *American Economic Review*, May 1971.

pay-cable households are not publicly available, and such data are regarded as proprietary by some operators.[15]

Eighth, as MacAvoy points out, the effect of eliminating the exclusivity, leapfrogging, and nonduplication rules on cable penetration has not been estimated. We have already noted that all but one of the studies of cable demand are based on data from operations prior to the imposition of these rules, so that the resulting estimates of penetration correspond to an unregulated environment. The proportion of an independent station's programming that must be blacked out because of the exclusivity restriction, both with and without substitution, has been calculated in another study by Park.[16] In the top fifty markets, signals are blacked out about 40 percent of the time. A rough indirect estimate of the effect on cable demand of lifting these restrictions could be constructed by using viewing-share data for blacked-out programming in the home station market to increase the "effective" number of imported signals in the cable-penetration equation.

In order to determine more reliably the quantitative effects of distant signals, local origination, and pay television, one could undertake another econometric study of cable penetration. Such an effort should combine a critical appraisal of the specification and the method of estimating the penetration equation with a new set of data on individual households, carefully constructed from a sample survey of a series of cable markets selected to provide independent variation in the key market variables of interest. Although careful design and data collection can undoubtedly improve somewhat the quality of results obtained heretofore, it will still not be possible to estimate the effect of relaxing specific rules questioned by MacAvoy. This is either because the natural experiment (relaxing exclusivity rules while retaining nonduplication) has not occurred or because that experiment (in the case of leapfrogging restrictions, which are now abolished) is too recent to have provided meaningful data.

---

[15] A research project into the demand for pay cable would have to be based either on acquiring system-by-system statistics on a confidential basis or on collecting original data through household surveys. Either type of data could be readily incorporated into a new cable-penetration study. The pay-cable study would systematically compile data on the availability, rates, and programming available on pay systems; the major determinants of pay penetration; and the evidence, if any, that pay programming increases the penetration of the basic cable system.

[16] R. E. Park, *The Exclusivity Provisions of the Federal Communications Commission's Cable Television Regulations* (Santa Monica, Calif.: Rand Corporation, 1972); summarized in *TV Communications*, August 1972.

Economic Viability of Cable Systems

Given the determinants of demand for cable television, the expected penetration in a specified set of market circumstances can be calculated. But will cable operators find it profitable to construct a system in a given area? To provide a method of answering this question several increasingly detailed models have been constructed to simulate the financial performance of cable systems—beginning with the Comanor-Mitchell study. The general structure of such models provides for capital-cost and operating-cost modules, a cable-demand module, and an accounting module that includes cash flow and internal rate-of-return outputs. All of these models are synthetic and eclectic, drawing their cost data for the specific components of a system from engineering specifications and field experience; no satisfactory data set exists from which to estimate econometric cost or production functions.[17]

The Comanor-Mitchell and Noll-Peck-McGowan studies—as well as studies by Mitchell-Smiley [18] and Crandall-Fray [19]—represent attempts to apply financial models of cable systems to a small number of "representative" circumstances that are assumed to typify important characteristics of market demand and cost—for example, signal quality, available local signals, income, underground requirements, and distance from imported signals. In contrast, the most recent models have been developed in case studies of specific markets and incorporate a considerable degree of city-specific detail for Dayton,[20] Washington,[21] Jacksonville,[22] and Baltimore and Cleveland.[23]

Several conclusions can be drawn from these models: First, although some differences with regard to individual cost components do exist among the models, such as the extent of scale economies in specific size ranges,[24] the models are in close agreement with regard to

---

[17] The models do include the costs of importing distant signals, contrary to MacAvoy's statement.

[18] B. M. Mitchell and R. H. Smiley, "Cable, Cities, and Copyrights," *Bell Journal of Economics and Management Science*, vol. 5 (Spring 1974).

[19] Crandall and Fray, "Prophecy of Doom for Cable Television."

[20] L. L. Johnson et al., *Cable Communications in the Dayton Miami Valley: Basic Report* (Santa Monica, Calif.: Rand Corporation, 1972).

[21] Mitre Corporation and Office of Telecommunications Policy, *Cable Television Financial Performance Model Description and Detailed Flow Diagram* (Bedford, Mass.: Mitre Corporation, 1974).

[22] Cable Television Information Center, *Cable Television Options for Jacksonville* (Washington, D.C.: Urban Institute, 1972).

[23] See Hopkins Cable Project study.

[24] See the studies by Comanor-Mitchell, Crandall-Fray, and Noll-Peck-McGowan.

the cost of delivering standard cable services. Local circumstances can be the cause of substantial cost variability. However, at prevailing monthly subscription rates both the market-characteristics models and the case-studies models arrive at break-even penetration rates in the 30 percent to 40 percent range for urban areas in the top 100 markets. If demand is approximately unit-price elastic in the relevant range, then systems that are estimated not to be viable at prevailing fees will tend to be unprofitable at higher prices.

Second, regulatory requirements that increase costs—such as copyright fees, two-way capability, commercial substitution, and cross-subsidies—can have important effects in making marginal systems not viable.[25]

Third, differences in the financial assessments for urban systems reached by the several studies stem primarily from variation in assumed penetration rates and, specifically, in the success that local origination and distant independents are assumed to have in boosting demand above levels predicted by the econometric demand models. For example, the Rand study by L. L. Johnson and others found a Dayton cable system unprofitable at a predicted 30 percent penetration, but viable if penetration were to achieve 40 percent as the result of substantial local origination and two-way services. The Hopkins Project study assumed one channel each for pay and for movies as well as eight for local origination and predicted 24 percent penetration for a Cleveland system; however, 31 percent to 33 percent would be needed to break even. Noll-Peck-McGowan estimated that by importing four strong independent signals large urban systems can achieve 59 percent penetration and can earn a 40 percent return on book value of investment.[26]

Fourth, restrictions on the product that cable systems can offer have the unambiguous effect of reducing expected profitability. We have noted the varied estimates of the effect of adding (or restricting) the number of distant independent signals. Restrictions on the nature of the imported signals—in the form of leapfrogging and exclusivity rules—either reduce the hours of imported programming the cable

---

[25] See Mitchell and Smiley, "Cable, Cities, and Copyrights," for an assessment of the effect of imposing such requirements on cable systems. A copyright bill has recently been enacted. The act provides for a compulsory license to import those distant signals currently permitted by the FCC. To obtain this license a cable system must pay a percentage of its revenue into a pool to be divided among the owners of copyrighted programming. The percentage is based on the number of signals imported.

[26] The estimate was for systems in middle-income neighborhoods requiring little or no underground construction.

system can offer or require the system to bear the costs of importing several signals in order to have a complete lineup of signals. Restrictions on the type of programming permitted on pay channels represents a further limitation on cable revenues and possibly on penetration.

Market circumstances and signal and program restrictions will determine the expected penetration of a cable system. Analysis of the financial viability of the cable system can then determine what subareas of individual markets will be profitable. When combined, these two components will predict the extent of cable penetration in a given television market, and, when aggregated, they will give a prediction for the United States.

Audience Diversion

How much audience will cable divert from local television stations? The diversion may be to distant signals, to pay movies, to cable origination, or to other services that cable systems may provide in the future.

Answers to questions about audience diversion require three kinds of information: (1) What sort of viewing options will be available on cable systems? (2) What fraction of households will subscribe to cable? (3) How will cable subscribers choose among viewing options? There are two studies that make explicit use of all three kinds of information to come up with estimates of audience diversion; these are discussed in the first section below. Item (1) is handled in the two studies simply by postulating an FCC policy that allows carriage of four distant independent signals and by assuming that cable systems would in fact carry those signals. Studies that attempt to answer item (2) have already been discussed above. Thus, our focus here is on the audience-behavior models necessary to answer (3). Besides the two studies discussed in the first section below, there are a number of other studies and scraps of data that shed further light on (3). These are reviewed in the second section. Conclusions that can be drawn from past research are listed in the succeeding section. A research project that would have a good chance of answering definitively some still unresolved questions about audience behavior is outlined in the fourth section. The concluding section tells why some other questions are probably unanswerable, at least in any definitive way.

Two Estimates of Audience Diversion. Park (1970) [27] and Noll-Peck-McGowan both make estimates of audience diversion because of cable carriage of four distant independent signals. They use very different methods, but their results are quite similar, at least in broad outline. Both sets of results are summarized in Table 10.

Park assumes that cable systems carry the following distant signals:

- any network signals necessary to provide three-network service,
- four very strong VHF independents in the top 100 markets, and
- three very strong VHF independents in smaller markets.

He does not assume any restriction on signal carriage because of leap-frogging or exclusivity rules.

His estimates of cable penetration range from 42 percent to 60 percent. Like the other early penetration estimates discussed above, these do not take account of the over-the-air reception difficulties that increase penetration in his sample. They are thus likely to overstate penetration in markets where reception is good.

Park does not estimate an audience-behavior equation. Rather, he simply assumes that total audience is not increased by the wider choices of options in cable, and that television households that receive a particular set of signals split their viewing among these signals in proportion to "attractiveness indices" that are based on the relative popularity with noncable subscribers of the signals in their home markets. The attractiveness index for a UHF station over-the-air is only about one-half of its index on the cable, because of over-the-air reception problems.

Park's assumptions about audience behavior seem plausible, but they are not subjected to any comprehensive empirical test. As a partial test, he compares his predictions of cable audiences' viewing of Los Angeles independent stations with actual data on Bakersfield and San Diego. He finds that his predictions are right on the mark for Bakersfield, but about 50 percent too high for San Diego.

---

[27] Park predicts the audience share for individual stations facing specified competition according to the following method: Audience is assumed to divide among signals in proportion to "attractiveness indices," a_i. If θ is the set of signals that can be received, the ith station's share $S_i = a_i / \Sigma_\theta a_i$. The a_i are calculated based on home-market audience shares. The relationship is applied separately to VHF households, UHF households, and cable households to estimate local audiences when cable imports distant signals. Further explanation is provided in Park, *Potential Impact of Cable Growth*, pp. 28-35; and ibid., *Cable Television and UHF Broadcasting* (Santa Monica, Calif.: Rand Corporation, 1971), pp. 3-25.

Table 10
TWO ESTIMATES OF AUDIENCE DIVERSION
TO DISTANT SIGNALS
(Four imported independent stations plus any missing network stations)

Number of Local Stations[a]				Percent Change in Audience for Station Class[b]			
NV	NU	IV	IU	NV	NU	IV	IU
Park (1970)[c]							
3		2+	0+	−13		−12	+24
3		1	0+	−16		−16	+24
3		0	1+	−18			+20
3		0	0	−16			
2	1			−22	+10		
1	2			−28	−3		
0	3				−17		
2	0			−34			
1	0			−48			
Noll, Peck, and McGowan (1973)							
3		3	1	±0		−16	+70
3		1	1	−4		−24	+100
3		1	0	−4		−24	
3		0	0	−8			
3		0	2	−8			+145
3		0	1	−8			+145
2		0	0	−29			
1		0	0	−50			

a NV: network VHF; NU: network UHF; IV: independent VHF; IU: independent UHF.

b These numbers are percentage change in *total* audience because of cable. That is, they take account of the fact that audience diversion occurs only in the fraction of the market that subscribes to cable.

c Plus sign following number of stations means "or more."

Sources: Park, *Potential Impact of Cable Growth* (1970), Table 6-8, p. 76; Noll, Peck, and McGowan, *Television Regulation* (1973), Table 6-4, p. 163.

His results on audience diversion are summarized in Table 10. In large markets, VHF stations lose about 15 percent of their audience to distant signals, while UHF stations gain 20 percent. In smaller markets, the audience loss is more severe, ranging up to a loss of almost 50 percent in one-station markets. These results apply to local audience only; independent stations carried as distant signals would pick up additional audience in other markets.

Noll-Peck-McGowan assume much the same complement of distant signals as does Park: affiliated stations as necessary to provide three-network service, and four independents (local UHF independents count against this total). Their cable-penetration equation, described in Table 7 above, yields estimates of penetration ranging from 48 percent to 85 percent. Like Park's equation, theirs neglects reception effects, so the estimates are probably overstated for markets with good reception.

Noll-Peck-McGowan estimate, using market data, two equations that predict individual station ratings as a function of the competition they face. The results, which are summarized in Table 11, are

Table 11
NOLL-PECK-McGOWAN: PREDICTIONS OF AUDIENCE DIVERSION

Dependent variables	Average number of homes viewing a particular station during prime time divided by the total number of television homes in the station's ADI (area of dominant influence).	
Samples	Network equation: random sample of 65 affiliates, approximately equally divided among networks. (1967)	
	Independent equation: all 46 stations for which data available. (1967)	
Method	Regression	
Explanatory variables	*Network equation*	*Independent equation*
UHF dummy	−, insignificant	−, significant
Competing affiliates	−, significant	—
Competing independents	−, insignificant	+, insignificant coefficient replaced by (−, insignificant) coefficient from network equation

Implied total market audience, in markets with:[a]	
1 NV	42–45 percent
2 NV	55–58 percent
3 NV	60 percent
3 NV, 1 IV	61 percent
3 NV, 4 IV	65 percent

[a] NV: network VHF; IV: independent VHF.
Source: Noll, Peck, and McGowan, *Television Regulation*, pp. 299-301.

not entirely satisfactory. The estimated effect of independent signals on network station audiences is negative as expected, but completely insignificant (t static about .5). The estimated effect of independents on other independents is even worse—positive, but insignificant. But since the effect of imported independents on local audiences is precisely the issue, Noll-Peck-McGowan have little choice but to make use of the estimates they get. What they do is to assume that the negative coefficient estimated in the network equation applies in the independent equation as well. As they point out, this probably overstates the effect of independents on other independents. In any event, Noll-Peck-McGowan also look at the Los Angeles independents' shares on cable in San Diego and Bakersfield, and they decide that these are roughly consistent with their estimates.

The Noll-Peck-McGowan estimates are shown in the bottom part of Table 10. They show less reduction in network station shares in large markets than does Park, and more help for UHF independents, but the overall pattern is remarkably similar. This coincidence of results from quite disparate studies lends some credence to both.

Other Studies of Audience Behavior. Statistical studies of audience behavior have also been performed by R. W. Crandall and by R. E. Park, L. L. Johnson, and B. Fishman. The studies are summarized in Tables 12 through 14.

These studies help to resolve a discrepancy between conventional wisdom, which holds that total viewing is insensitive to the number of viewing options, and Noll-Peck-McGowan, whose equations show prime-time audience increasing from between 42 percent and 45 percent with one network signal to 60 percent with three networks and 65 percent with three networks plus four independents. The other studies indicate that viewing is less sensitive to the number of options, especially to the addition of the second network.

The Noll-Peck-McGowan estimates have an error-in-variables problem in their data that may bias their results. By ignoring overlapping signals from adjacent markets, Noll-Peck-McGowan underestimate both the actual number of viewing options and the total audience. These errors are correlated, since both are larger in markets where overlapping signals are more important. If markets with one or two networks are more likely to have overlapping signals, Noll-Peck-McGowan will overestimate the audience gain because of more networks, since their procedure makes these markets look as though they have fewer options and smaller audiences than they really do.

Table 12
CRANDALL STUDY: MARKETS EQUATIONS

Reference	R. W. Crandall, "The Economic Case for a Fourth Commercial Network," *Public Policy*, Fall 1974, pp. 517–22.
Dependent variables	Somewhat unclear: "the percentage of television homes viewing during prime time." [Is this ADI HUT?] "Total number of quarter hours watched by average household in one week." [In the ADI?]
Sample	All 207 ARB markets. (1970)
Method	Regression
Explanatory variables	
Number of affiliates	+, significant
Number of independents	±, insignificant
Number of noncommercial	±, insignificant
Central time zone	+, significant for prime time percent −, significant for weekly quarter hours
Number of TV households	+, insignificant
Predicted audience outside central time zone	
1 NV market	55 percent
2 NV markets	56 percent
3 NV markets	57 percent
Cable homes	53 percent

The other studies show audience increasing between two and six percentage points as options go from one to three network signals. These studies also provide biased estimates of the effect of more networks, with the bias operating in the opposite direction. In these studies, all stations satisfying a particular criterion (for example, the area is within the station's Grade B contour or the station achieves some minimum audience rating) are counted as full competitors. But, in fact, not all such signals can be received with adequate clarity, so that the Crandall and the Park-Johnson-Fishman procedures overcount the competition offered by these signals. This leads to an underestimate of the contribution of additional signals to total viewing, by the same reasoning applied to the Noll-Peck-McGowan analysis.

Table 13

CRANDALL STUDY: COUNTY EQUATIONS

Reference	R. W. Crandall, "The Economic Case for a Fourth Commercial Network," *Public Policy*, Fall 1974, pp. 522–25.
Dependent variables	Average percentage of homes watching television during prime time in each county.
	Total quarter hours watched per week by average household in county.
Samples	(1) All 262 ARB counties in which only local signals are reported viewed.
	(2) All 235 such counties with no cable subscribers. (1970–71)
Method	Regression

Explanatory variables
Number of affiliates +, significant
Number of independents ±, insignificant
Number of noncommercials ±, insignificant
Central time zone +, significant for prime time percent
−, insignificant for weekly quarter hours
Number of TV households +, significant

Predicted market audience
in counties outside central
time zone with 1,000 TV
households
1 NV market 45 percent
2 NV markets 49 percent
3 NV markets 51 percent

Park-Johnson-Fishman find a significant further viewing increase (one percentage point) in markets with independent signals as well, but Crandall does not. The Park-Johnson-Fishman result comes from classifying all markets with VHF independents as identical, regardless of the number of these stations. Since all but three such markets have but one station, the dominant influence in the Park-Johnson-Fishman results is the single-independent market. Hence, the Park-Johnson-Fishman and the Noll-Peck-McGowan estimates of the audience effect of independents are roughly consistent, both being near 1 percent.

Table 14

PARK-JOHNSON-FISHMAN STUDY

Reference	R. E. Park, L. L. Johnson, and B. Fishman, *Protecting the Growth of Television Broadcasting: The Implications for Spectrum Use* (Santa Monica, Calif.: Rand Corporation, 1976), pp. 195–200.
Dependent variable	Average percentage of homes watching television during prime time in each county.
Samples	(1) All 3,094 ARB counties in 48 contiguous states.
	(2) All 2,672 rural counties.
Method	Analysis of variance

Total audience (percent)

1 NV market	54	} insignificant difference
2 NV markets	56	} significant difference
3 NV markets	58	} significant difference
3 NV, 1 or more IV markets	59	

Results are the same for total and rural samples, but vary slightly depending on how stations are counted.

For markets with two or more networks, all of the studies are in rough agreement with regard to the effect of additional signals. In particular, in markets with three networks, all studies conclude that more viewing options of either type would have only a minor effect on the total size of the audience.

S. M. Besen and B. M. Mitchell find that dramatically different programming (for example, the Watergate hearings) can tempt a large number of new viewers.[28] Additional conventional programming is not sufficiently different to have the same effect.

There are also some scattered data on cable audiences for imported independent signals that tend to support one assumption and one conclusion on which Park (1970) and Noll-Peck-McGowan agree. Both assume that independent stations carried into distant markets by cable will attract as much audience as they would locally against equivalent competition. Cases 1–3 in Table 15 support the assump-

[28] S. M. Besen and B. M. Mitchell, *Watergate and Television: An Economic Analysis* (Santa Monica, Calif.: Rand Corporation, 1975); a revised version appears in *Communication Research*, July 1976.

Table 15

AUDIENCE SHARES PER STATION FOR INDEPENDENT STATIONS OFF-THE-AIR AND ON CABLE

Case	Stations	Where Watched	Prime Time Off-air	Prime Time Cable	Fringe Off-air	Fringe Cable	All Day Off-air	All Day Cable
1	4 Los Angeles VHF independents	Bakersfield	1	7	1	10	1	8
2	3 Los Angeles VHF independents	San Diego	1	3	2	8	1	5
3	Average of 1.47 independents	82 counties outside B contour	0	5	2	12	1	6
4	3 Philadelphia UHF independents	Off-air: Philadelphia Cable: Lebanon	5	6	6	11	—	—
5	1 Pittsburgh UHF independent	Allegheny County	—	—	11	11	5	6
6	3 Philadelphia UHF independents	Northampton County	—	—	2	4	1	3
7	1 San Diego UHF independent	San Diego County	—	—	2	4	1	2

Note: The reader should be careful in comparing the numbers in this table with the numbers in other tables in this chapter. There are two important differences in meaning: (1) The numbers in Tables 11-14 are total "ratings," or percentages of *potential* audience that actually watches. The numbers in this table are "shares," or percentages of *actual* audience. (2) The numbers in Table 10 refer to changes in *total* audience. The numbers in these columns refer to *cable* audience alone. For example, Case 1 shows that Bakersfield stations lose 27 percent of their prime-time audience to Los Angeles independents in cable households. But if only half of the television households in Bakersfield subscribe to cable, this is only a 13.5 percent loss in total audience.

Sources: Cases 1 and 2: FCC Staff Report, Research Branch, Broadcast Bureau, "The Economics of the TV-CATV Interface," Washington, D.C., July 15, 1970. Case 3: Statistical Research, Inc., *The Potential Impact of CATV on Television Stations*, Appendix E to National Association of Broadcasters filing in FCC Docket 18397-A, Fall 1970. Case 4: Kaiser Broadcasting Corporation, Filing in FCC Docket 18397-A, December 4, 1970, reproduced in Park, *Cable Television and UHF Broadcasting*. Cases 5-7: American Broadcasting Company, Filing in FCC Docket 18397-A, December 1970, reproduced in Park, *Cable Television and UHF Broadcasting*.

tion. All show small audiences for independents carried well outside their home market (as indicated by the low off-the-air shares).

Cases 4–7 lend support to the result that cable will tend to help UHF independents. Case 4 shows the three Philadelphia UHF stations doing better on the cable than over-the-air in their own metropolitan area, and Case 6 shows them doubling their audience on the cable in the outlying portion of the Philadelphia ADI (area of dominant influence). Cases 5 and 7 compare UHF independent audiences on the cable and over-the-air in the center of their markets. Cable audiences are at least as large as over-the-air audiences in all cases. Cases 3 through 7 all come from studies financed by the broadcast industry. It appears that industry studies are in substantial agreement with independent researchers on the amount of audience diversion on the cable.

Conclusions from Past Research. The following conclusions are justified (though perhaps not definitively established) by past research.

1. Cable carrying four distant independent signals would not divert more than 15 percent to 20 percent of the audience of VHF stations in large markets, even assuming unrealistically high penetration (on the order of 50 percent).

2. Audience diversion to distant signals on cable in smaller (one- and two-station) markets will be (and is) substantial. The greater diversion in these markets is because of greater cable penetration, and because of importation of distant network affiliates. These are mostly in the group below the top 100 markets, where cable operated without restriction before 1972. It is worth noting that even in these markets the number of stations has not declined in the face of increasing cable penetration.

3. Cable helps UHF independents in their own markets, because improved reception outweighs audience diversion to competing signals. This effect is greater at the edges of the market than it is in the center.

4. The size of the total audience is mildly responsive to the number of viewing options, at least going from one to three network signals.

A Research Project Using New Data. Previous studies of audience diversion have had to get at cable viewer behavior indirectly, either by assumption (Park) or by estimates based on off-the-air viewer behavior (Noll-Peck-McGowan). The data now exist for a direct investigation. Data collected by the American Research Bureau (Arbitron) for each U.S. county that measure the audience of all significantly viewed stations are now available separately for households viewing on cable and those viewing off-the-air. The data are collected by placing viewing diaries with sample households for two-week periods two or more times per year. Three years' worth of data are now available, for markets in which cable penetration exceeds 10 percent.

The availability of separate measurements on cable and noncable subscribers in the same market offers an important opportunity for advancing an understanding of the demand for television, and, in particular, an understanding of the impact of cable on the audiences of local broadcast stations. Previous studies have used county audience data on all households to explain total viewing in a market or county as a function of the number and types of signals available. In contrast, the proposed study would disaggregate total viewing both by type of station (for example, local network, distant network, local independent, distant independent, public) and by type of viewer (cable and noncable). In this way the effect of adding cable viewing options and the resulting impact on local stations' audiences can be directly measured.[29]

Some of the major questions that such research might answer definitively are:

- Are cable subscribers persons who watch a greater-than-average amount of television?
- Does cable increase the total amount of viewing?
- How much does cable reduce the viewing of each type of local station?
- How much new audience does cable provide for imported stations, classified by type and by distance from the local market?

Although the proposed study would add to the detail and precision of our knowledge of viewer behavior, we do not expect that it would produce results that conflict with the conclusions listed above.

---

[29] This study is similar to that proposed by MacAvoy; see p. 39 above. Several methodologies could usefully be applied in such a study, including: (1) Reestimating the total audience equations from earlier studies, but using cable and noncable households as separate observations; (2) Developing the audience-split models suggested by Park's and Besen's work, in which departure from equal shares of the total audience is a function of the programming and the signal quality of all alternatives; (3) Generalizing the Besen-Mitchell method of estimating subgroups of viewers who have the same preference orderings for program alternatives to incorporate the fact that the observed cable viewers are, by self-selection, those with the greatest willingness to pay for cable options.

Some Questions that Cannot Be Answered Definitively. Past research tells us a fair amount about audience diversion to distant signals. Further, new research, using newly available data on cable viewing, promises to add to the detail, confidence, and precision of our knowledge about the same subject—perhaps even to "settle the debate." Questions about diversion to pay television and other new services, however, are much harder to answer.[30] The reason, of course, is that we have very little experience and almost no data to go on.

In fact, one can make much the same point about possible future developments in distant signals. For example, a big-city independent station that is carried on many cable systems may act like a mini-network, substantially upgrading the attractiveness of its programming. What would be the consequent audience diversion? The question cannot be answered definitively, because we do not yet have any experience with such super stations.[31]

We can hope to make more-or-less definitive, research-based statements about things we have experienced—for example, audience diversion to distant signals that are imported subject to past and present restrictions on leapfrogging and exclusivity. But the farther one projects into regions untouched by existing data, the less likely it is that the projections will settle any debates.

Economic Viability of Stations

Given the level of cable penetration and its effects on audience diversion, one is in a position to attempt to estimate the effects of the introduction of cable on the viability of broadcast stations. In fact, four studies have attempted in different ways to deal with this issue. In three of the studies (Fisher-Ferrall,[32] FCC staff,[33] and Park [34]) the

---

[30] MacAvoy has suggested that further analysis of the Hartford experiments would be useful; see p. 43 above. This is almost certainly fruitless because: (1) Hartford had no access to programming unless it was produced for another purpose since it was such a small operation; (2) nearly all offerings were movies, and the value of analyzing the success of different types of movies is dubious; (3) all the movies were quite recent when shown, so one cannot even investigate the importance of the three-to-ten-year rule; and (4) the number of programs in each nonmovie category was so small that it would not support statistical analysis.

[31] One could make fairly convincing upper bound calculations, however, by treating the super stations as the equivalent of a network signal.

[32] F. M. Fisher, and F. E. Ferrall, Jr., et al., "CATV Systems and Local TV Station Audiences," *Quarterly Journal of Economics*, May 1966.

[33] FCC, Broadcast Bureau, Research Branch, "The Economics of the TV-CATV Interface," FCC Staff Report, July 15, 1970.

[34] Park, *Potential Impact of Cable Growth.*

effect of audience fragmentation on the revenues of broadcast systems are examined. Then, using assumptions about the effect on station costs, the effect on station viability is assessed. In a more recent study, Park, Johnson, and Fishman deal with the effect of cable penetration directly on station viability.[35]

The studies by Fisher-Ferrall and Park estimate a relationship between a station's revenues and its prime-time audience. Fisher-Ferrall find that a reduction of one viewer resulted in a reduction in station revenues of about twenty-seven dollars in 1963. Park finds, by reestimating the equation using 1968 data, that this had risen to about forty-three dollars. Park also argues that, because of the differing ways in which network affiliates and independent stations receive their revenues, the revenue-audience relationship should differ between them, and he finds this hypothesis consistent with the data. Park finds that the revenue-audience relationship is quadratic, indicating that the additional revenue per viewer declines once audiences exceed a certain point. Finally, the FCC staff in their study of viability simply assume that the decline in revenues will be proportional to the decline in audience that they estimate.

To do a study of the effect of cable penetration on station profitability requires an assessment of its effects on costs as well as on revenues. To our knowledge, such a study has not been done. Fisher-Ferrall, after estimating the effect of audience reduction on station revenues, simply assert that many stations will not be able to withstand the measured revenue reduction and still stay on the air. Both Park and the FCC staff assume that station costs will not change as their audience and revenues decline. This assumption is unrealistic; the effect of making it biases their findings on viability toward the conclusion that stations will fail.[36]

The FCC staff study concludes that the estimated audience (and therefore revenue) reduction will be larger than the profit margins of many of the affected stations, and, since costs are not assumed to change, these stations will become unprofitable. Park carries out a more detailed analysis using the same FCC profit data of the effect of the estimated reduction in revenue on station viability. He finds that there are small reductions in the number of profitable stations

---

[35] R. E. Park, L. L. Johnson, and B. Fishman, *Projecting the Growth of Television Broadcasting: The Implications for Spectrum Use* (Santa Monica, Calif.: Rand Corporation, 1976).

[36] Even stations which cannot survive on their own can become the satellites of other stations and simply retransmit their programming. This activity can be carried out at very low cost. If a substantial number of households do not subscribe to cable, this will be a profitable activity.

in those markets which already have at least one independent station. In markets where there are three network affiliates but no independent station, there is a somewhat larger effect on the number of stations that are profitable. By far the largest impact on the number of profitable stations is in those markets which presently do not have full network service.

Aside from the use of the assumption that costs do not change as station audience and revenues decline, the FCC staff and Park studies suffer from the common difficulty of having to rely on existing FCC profit data. A recent study by Park-Johnson-Fishman points out that there are numerous shortcomings in these data, the principal one being that there is no common set of accounting rules which guide stations in preparing their reports to the FCC. As a result, quite wide divergences of operating results are measured for stations in essentially the same operating circumstances. Although some differences might be expected because of differences in market characteristics and managerial performance, the measured differences are much larger than could be accounted for in this way. As a result, Park-Johnson-Fishman are led to abandon any attempt to use these data to assess the factors which determine station profitability. Further confirmation of the limited usefulness of these data is the fact that some stations report operating for many years with consistent accounting losses. This would be possible only if (1) reported profits are meaningless, or (2) station owners and the investment community persisted in expecting better days ahead. What this suggests is that there might be a substantial payoff to the adoption of a uniform system of accounts to guide stations in making their reports to the commission. To be useful, not only must the reports be done under a uniform format, but the figure reported as profits must be the same as that relied upon by station owners in deciding whether or not to continue to operate.[37]

Park-Johnson-Fishman attempt to explain directly the number of stations that operate in a given market, thus avoiding the need to rely on the FCC profitability data. One of the factors they assess is the effect on station viability of cable television. In their base-case projections they find that, in 1990, there will be ten fewer stations than there would have been in the absence of cable but seventy-nine more stations than there are at present. In their "worst case" projections, those that maximize the likely effect of cable on broadcasters,

---

[37] This suggests why it would be inadvisable to pursue the suggestion to make detailed station-by-station projections of revenues and costs for the next five to ten years until such a change in reporting methods occurs.

they find that cable will result in forty-one fewer stations than otherwise, but that there will still be forty-six more stations than there are at present. The analysis is confined to the top 100 markets.

Although the results of the above studies, especially those that rely on FCC profitability data in assessing viability, are certainly not definitive, the results they obtain are plausible. The largest impact of cable development on broadcasters occurs in markets with relatively poor over-the-air service. These are markets which have only one or two stations operating over the air, almost all of which are markets below the top 100.

Cable affects broadcast audience, revenues, and presumably profitability because the network affiliates imported into these markets receive a very large share of the audience. It should also be pointed out here that, for markets below the top 100, FCC protection has been weaker than for any other group of stations. The rules which prevailed before 1972 afforded them no protection at all. The existing rules define minimum service in these markets to be of a lower quality than in markets in the top 100. Most of these markets, however, have fewer than three affiliates, and the impact of importing additional network affiliates affects these stations far more than does the importation of independent stations into those markets which already have full network service. Moreover, only the top 100 markets are afforded exclusivity protection. (This is not to argue that stations in these markets should have received more protection than they have but, instead, to make the point that the most vulnerable stations have been afforded the least protection.)

Despite this threat from cable, however, the number of stations in these markets did not decline in the decade before 1972. Compared to the impact of cable in markets below the top 100, its impact elsewhere is quite small.[38] In markets with full network service, cable penetration will be lower and audience fragmentation more limited than in the one- and two-station markets. The most vulnerable stations in the larger markets are, not surprisingly, the independents.[39]

In fact, the forecasts in these studies of the deregulation have largely been borne out. There have been few, if any, reductions in

---

[38] Thus, contrary to the statement by MacAvoy, the existing studies do deal with the question of which stations will most strongly be affected by unlimited importation.

[39] Our ability to assess the effects of cable importation on local educational stations is even more problematic since, until recently, audience data for these stations were not collected systematically. Moreover, the factors which explain the viability of these stations are much more difficult to model than are the factors which determine the viability of commercial stations.

the number of stations in mid-size markets during the period since the relaxation in the cable rules in 1972, contrary to the impression left in the MacAvoy memorandum.[40]

In addition to the improvement in data collection suggested above, one other possible study might also contribute to an understanding of the effect of cable on station viability. This is to analyze the data on the prices at which television stations are exchanged in order to determine whether these prices are affected by cable penetration. While these data are not without their difficulties (one must control for differences in equipment replacement costs and deal with the fact that frequently the transactions are complicated ones involving multiple properties), they do provide one place in which, without relying on existing FCC data, the effect of cable penetration on viability can be assessed. Approximately 250 stations have changed hands since 1960, thus providing a large data base for such a study.

For the reasons suggested above in discussing the feasibility of measuring the effect of pay television on penetration and on audience diversion, it is impossible to determine the effects on broadcasters. Not only is there no nonproprietary data on the audience for pay television but we cannot possibly know what a full-blown pay-television system will be able to provide in the way of programming. While there have been a number of analyses of the possible effects of pay television (for example, Owen-Beebe-Manning and Noll-Peck-McGowan), the testing of the hypotheses that these studies suggest awaits the day when data on pay television become available and when pay television is allowed to provide, and does provide, programming different from its current fare.

Programming

Deregulation of cable television (distant signals and pay television will almost certainly change the number, type, and quality of programs offered to the public, and it will do so differentially in markets with different characteristics. The fundamental policy research question is to identify the gains and the losses to particular consumer groups from these changes.

Among the particular questions that arise in this context are the following: (1) Does local origination on cable affect penetration or

---

[40] Nevertheless, detailed analysis of the question whether stations failed as a result of the rule changes is not available. This question might be addressed by a comparative analysis of stations which failed during this period with similarly situated stations which survived.

attract enough audience to be commercially viable, and, concomitantly, what amount of local origination can be expected of cable systems? (2) What is the effect of cable penetration on the programming policies of individual broadcast stations and networks, including the specific effects on locally originated broadcast programming? (3) To what degree will cable pay television "siphon" programming from advertiser-supported over-the-air television? (4) What will happen to the supply of television programming if cable and pay increase the demand for it?

Several of the major studies of demand for cable television included a variable for local origination by the cable system. The studies uniformly find no effect from local origination except for the Hopkins Project study which finds a very small, positive effect on penetration. The FCC has found it appropriate to require cable systems to originate programming. It is not surprising that there should be little demand for local origination on cable systems, given the experience of broadcast stations with much larger potential audiences; with the exception of local news, such programming is not profitable. Since programming cost is independent of audience size, access to a small potential audience ensures that cable-originated local programming will be even less profitable.

It is conceivable, however, that the coverage of cable systems may be sufficiently localized, compared to the broader coverage of television stations, so that neighborhood programming and neighborhood advertising could be profitable at sufficiently high penetration rates. It does not seem possible to do useful research on this issue, since such programming and advertising do not yet exist.

As MacAvoy points out, increased competition from cable will cause broadcasters to alter their programming strategies. Broadcasters have argued that the main effect will be to reduce the quantity and the quality of local programming, which is provided at financial loss as a public service. No empirical research has been done on the question of what these altered strategies would produce.[41] There is,

---

[41] One could examine changes in programming by local stations over time, as cable penetration grew as a function of the amount of audience diverted by the cable system and by the nature of the services on the cable that diverted the audience. This would require a monumental data-gathering effort. One would have to collect programming logs for all the stations and note other services on a cable system over several years. One would also have to construct meaningful program categories, a task that has thus far proved difficult except in rare, unique instances (for example, the Watergate hearings). The scientific value of results here would be quite high; the costs would also be significant, and the results would probably be a few years in coming. And even then, the policy significance of the results would be controversial, since no big cities have the

Table 16

COMMERCIAL RADIO STATIONS IN OPERATION

Year	AM	FM	Total
1945	884	46	930
1950	2,086	733	2,819
1955	2,669	552	3,221
1960	3,456	678	4,134
1965	4,012	1,270	5,282
1970	4,319	2,184	6,503
1975	4,472	2,806	7,278

however, an extensive theoretical literature on program patterns under various industry structures.[42] These studies indicate that an increase in the number of competitors is likely to lead to an increase in diversity, and (with pay television) to an increase in consumer welfare. Since it is difficult to measure the demand for a product whose price is zero, empirical estimates are unlikely to be possible. There is, however, an extensive literature on the social value of television programming.[43] This literature is entirely noneconomic in content, and there is very little that we can say about it.

It may be possible to gain some qualitative insights into the effects of cable deregulation on programming by looking at the history of radio broadcasting. Radio has gone through the experience both of increased numbers of channels (see Table 16) and of competition from a new technology (television). Radio programming has changed as a result, largely by becoming more specialized. But it is

kind of cable penetration that cable advocates hope and expect will eventually emerge. Since broadcasters have enormous rents and, generally speaking, have thus far been almost totally unaffected by cable, it is likely that the effect of cable on programming to date bears little relation to the response after massive cable development.

[42] For example, P. O. Steiner, "Program Patterns and Preferences and the Workability of Competition in Radio Broadcasting," *Quarterly Journal of Economics*, May 1954; B. M. Owen, W. G. Manning, Jr., and J. Beebe, *Television Economics* (Lexington, Mass.: Lexington Books, 1974), chapter 3; and M. Spence and B. M. Owen, "Television Programming, Monopolistic Competition, and Welfare," in B. M. Owen, *Economics and Freedom of Expression: Media Structure and the First Amendment* (Cambridge, Mass.: Ballinger Publishing Co., 1975).

[43] See, for example, Douglass Cater and Richard Adler, eds., *Television as a Social Force: New Approaches to TV Criticism* (New York: Praeger Publishers, 1975).

problematical how much inferential value such a study would have in predicting television programming changes.

The effect of pay-television operations on television broadcast programming has not been extensively analyzed because there are insufficient data. Broadcasters argue that there will be extensive "siphoning" of existing broadcast programming to pay operations. Again, the difficulty is that it is hard to measure demand for a product whose price is zero. However, a number of arguments are relevant. First, the pattern of exhibition of motion pictures which now exists (several releases to movie theaters followed by several releases to television) suggests that pay television would be only an additional step in the release pattern. Indeed, to the extent that release to pay cable with high-penetration substitutes for theatrical release, movies may come to broadcast television sooner than they now do. Extensive siphoning of series seems most improbable even under optimistic assumptions about the growth of the pay audience.[44]

One fear about pay television is that very high prices will be charged, with resulting adverse effects on the poor. If the *current* system were financed entirely by pay rather than by advertising, the average cost would be about three cents per viewer hour, plus transactions costs. Given competition and free entry, prices would presumably be in the same range. The redistributional effects do not seem worth worrying about. Of course, a truly national, multichannel, high-penetration pay-cable system (such as might exist in twenty years) would have a different cost and demand structure. But, in any event, the continued demand for large audiences by advertisers and the elastic supply of programming material suggest that the existence of pay television is unlikely to reduce the supply of large audiences to advertiser-supported broadcast programming.

The growth of additional channels as a result of pay television implies an increased demand for programming, whether advertiser-supported or paid for by viewers. It is therefore of interest to examine the elasticity of program supply. The major study of this point is in Owen-Beebe-Manning, where it is argued that the supply of programs is likely to be very elastic.[45] Crandall has argued that the major studios are able to restrict the supply of feature films, but this (if true) is because of their oligopolistic control of the "bottleneck" distribution stage. Presumably, cable and television networks eliminate this bottleneck by dealing directly with producers. The programming supply

---

[44] See the Noll-Peck-McGowan study.

[45] Owen, Manning, and Beebe, *Television Economics*, chapter 2.

industry is characterized by easy entry and specialization of rental factors of production, so that expansion of supply should not require substantial price increases. Sports are sometimes regarded as a special case because of the alleged inelasticity of supply of certain events. However, even extraordinary events are not in completely inelastic supply (for example, there was once no Super Bowl). Expansion in several major league sports, the advent of new televised sports (for example, tennis and soccer) and the growth of college football bowl games all suggest that supply inelasticities are easily exaggerated.

If the date were 1946, we would be asking whether to "deregulate" (actually, allow entry of) television broadcast stations. One of the issues would be the effect of television on radio programming. Will it drive radio stations off the air? Will it deprive nonviewers of local radio service? Will "The Shadow" be "siphoned" to television? Will poor people be able to afford television sets? Most of these questions were as unanswerable in 1946 as their analogues are today. It is difficult or impossible to predict the demand for and the diffusion of a new technology or a new product. Did the FCC err in allowing development of television, since it did not prove that radio stations and radio listeners would not be harmed, much less which would be harmed by how much? If the FCC had in 1946 adopted the burden of proof proposed by MacAvoy, the cable television problem would not be with us today, for we would still be without television.

The FCC has consistently emphasized the social utility of locally originated programming, and Congress has at least implicitly accepted this policy objective. In practice, the FCC has tolerated a perfunctory amount of local programming by broadcast stations,[46] and it has relieved cable operators of their origination obligations when these appeared onerous. Localism is a policy objective without much practical importance, even in the hands of its foremost proponent. Since research on the programming effects of cable deregulation could not be definitive even on the broadest level, it certainly could not settle the debate about local programming. But, whatever social and political value localism possesses, it would be strange to find policy makers giving up very much of it in order to protect the amount we now enjoy.

---

[46] Television stations average about 15 percent locally produced programming, the majority of it being local news. The local programs that are not news or sports have audiences so small as to be within the range of measurement error of the ratings services.

4

RECENT DEVELOPMENTS AND PREDICTIONS ABOUT THE FUTURE

Kenneth Robinson

The Domestic Council effort involving cable television "deregulation" did not have much force from June 1976 on. There was little agreement even within the DCRG about what should be done, and the feeling was widespread among the DCRG members that maybe it had bitten off more than it reasonably could chew. Some significant forward action did occur, however, as Congress enacted copyright reform legislation and as major portions of the FCC's cable television regulations were gutted by the appellate courts, partly because of Justice Department efforts. We shall review those important steps here.

Recent Developments

In late 1976, Congress finally enacted copyright reform legislation, thus concluding nearly a generation of sometimes vituperative debates. The new law overturned *Fortnightly* and other Supreme Court rulings,[1] for the first time extending copyright liability to include cable retransmission of over-the-air broadcast signals. Public Law 94-553 was signed into law by President Ford on October 19, 1976.

Under this new copyright law, cable systems are accorded a compulsory copyright license, permitting them to retransmit broadcast signals without the station's consent effective January 1, 1978. In return, virtually all cable systems are required to pay royalties, the actual amounts to be determined generally on a system-by-system basis. A complicated formula keyed to the number of "distant-signal equivalents" a given system may import determines the payments, except for the smallest CATV systems, which instead pay a fixed

[1] Fortnightly Corp. v. United Artists Television Corp., 392 U.S. 390, 400 (1968); TelePrompTer Corp. v. CBS, Inc., 415 U.S. 394, 405 (1974).

percentage of their revenues.[2] Theoretically at least, the imposition of this new copyright liability erodes the traditional FCC rationale for curbing CATV service availability, namely, that CATV constitutes "unfair competition" because it "pirates" programming only broadcasters were required to pay for.

Whether this erosion of the traditional rationale will have any practical effect, though, is another question. Section 801(b)(2)(B) of this very complex law could complicate efforts to relax existing FCC regulations. That section essentially codifies the FCC regulations regarding distant signals and program exclusivity as they existed on April 15, 1976. The law provides that, should these regulations be changed in the future, the statutory royalty fee schedule in the act may be revisited and the rates increased. This reopening of the fees question would pertain only with respect to the additional services CATV systems might be permitted to offer. The rates for services authorized as of 1976 are generally "grandfathered" in a fixed schedule.

The value of more distant signals and less FCC-mandated "blacking out" of retransmitted programming to both cable system operators and their subscribers is still an open question. Whether the gains that could be expected by further liberalizing these rules outweigh the obvious risks involved in reopening the matter of the fees payable, if only in part, is certainly not clear. Hence an effect of this particular provision in the 1976 act may be to freeze standard or basic CATV service offerings at 1976 levels, making it likely that so-called new and innovative services theoretically offerable by cable will be the only practical avenue for expanding a system's repertoire of offerings.

Although the effect of recent copyright legislation is unclear, the 1977 decision by the Court of Appeals for the District of Columbia Circuit that reversed the pay-cable rules should have at least a substantial psychological effect.[3] On March 25, 1977, the D.C. Circuit issued its ruling in the *Home Box Office* case, a consolidated appeal by many parties challenging the validity of the FCC's 1975 pay-cable rules. In 1975–1976 the commission had adopted rules that, in essence, barred the exhibition of feature films by cable systems on a per-program or per-channel pay basis if those movies were more than three years old. The pay-cable rules also barred pay-cable exhibition of specific sports events, such as the World Series, and placed extraordinary restraints on the number of ordinary season games pay cable might exhibit. The putative rationale underlying these restrictive

---

[2] See P.L. 94-553, section 111(d) and following.

[3] Home Box Office, Inc. v. FCC, No. 75-1280 et al., U.S. Court of Appeals for the District of Columbia Circuit, March 25, 1977.

FCC rules was the need to protect against "siphoning." As the commission saw things, pay-cable systems could potentially muster the buying power to outbid "free television" and the national networks for the most popular television fare. Once these programs were siphoned away, they would be made available only to those few who had access to cable services and who, in addition, could afford to pay extra amounts for them. Interestingly, the FCC, while maintaining that these popular programs must be available to free-television viewers, specifically declined to require the broadcasting industry actually to show them, citing First Amendment grounds.

The commission's pay-cable rules were challenged by the Justice Department and the cable industry as well as other parties, on jurisdictional, First Amendment, procedural, and competitive grounds. Established broadcast interests also challenged the commission's action, asserting that these highly restrictive regulations were still too weak. The FCC defended its actions in adopting curbs on pay cable and pay over-the-air television as legitimate exercises of its regulatory discretion. It argued that institutional expertise was sufficient to support virtually any restrictions it felt were desirable to protect the "public interest." The commission asserted that specific record evidence of actual harm was not required before imposing prior restraints on pay-cable services. It dismissed allegations of improper, ex parte contacts with both cable and broadcast interests in devising its pay rules, arguing that restrictions should not be placed on open-ended, general rule makings such as this, where frequent, informal contacts with affected industries allegedly were a valuable component of the commission's decision-making process.

The court of appeals rejected all of the FCC's arguments, squarely reversing it on virtually all points, in a 106-page, strongly critical opinion. The court ruled that the pay-cable rules as adopted far exceeded the commission's jurisdiction over cable, which remained ancillary at best. They violated the First Amendment in that they were grossly overbroad and not based upon any credible evidence that restrictions were required in the first place. The court also characterized the commission's illegal, ex parte procedures as "intolerable."[4] The court laid out a number of alternative grounds to support its ultimate decision: that the rules were void and should be dismissed. So many alternative grounds were spelled out, moreover, as to undermine seriously any expectation on the FCC's part that the D.C. Circuit could be reversed on the result should the Supreme Court take any further appeal.

---

[4] Ibid., slip opinion, p. 89.

The *Home Box Office* case should provide grist for many legal and academic literature mills, assuming, as seems likely, the case is upheld on appeal. Three key features of the opinion, however, seem particularly important. First, the court held that the FCC erred in adopting program content restrictions for cable because in the commission's view, CATV was "just like television." The court reasoned as follows:

> The First Amendment theory espoused in *National Broadcasting Co.* and *Red Lion Broadcasting* cannot be directly applied to cable television since an essential precondition of that theory—physical interference and scarcity requiring an umpiring role for the government—is absent.[5]

In terms of what, if any, content controls and limitations might be permissible, the court concluded that cable television systems should stand toward government not as broadcast licensees but as newspapers. Under long-established Supreme Court doctrine, government power to control the content of newspapers, of course, is practically nonexistent.[6]

Second, the court concluded that the FCC's position that cable television should be a secondary or supplemental service to television broadcasting was in error.

> (T)he Commission has in no way justified its position that cable television must be a supplement to, rather than an equal of, broadcast television. Such an artificial narrowing of the scope of the regulatory problem is itself arbitrary and capricious and is ground for reversal.[7]

Since perhaps the major underpinning of many of the current restrictions the FCC has imposed on cable television service is the assertion that CATV should be a "second-class citizen," this finding by the court may put into question the continuing validity of other FCC cable regulations, such as those relating to distant-signal importation.

Third, the D.C. Circuit largely dismissed the theory of presumptive harm upon which the FCC's pay-cable rules were based. The court noted, as had the Justice Department repeatedly before the commission, that there was no factual predicate for the assumption on the FCC's part that, absent restrictions on cable, "free television" would be adversely affected. Furthermore, the court concluded, the

---

[5] Ibid., pp. 69-70.
[6] See, for example, Miami Herald Publishing Co. v. Tornillo, 418 U.S. 241 (1974); Grosjean v. American Press Co., 297 U.S. 233 (1936).
[7] Home Box Office, Inc. v. FCC, slip opinion, p. 52.

commission had failed to demonstrate how the harms it had hypothe-
sized would adversely affect the public, pointing out that television
industry profit levels and service to the viewing public had not been
shown to be causally related on the record.[8] This rejection of the
FCC's doctrine of presumptive harm, obviously, has application in a
number of other regulated industry contexts.

As a legal proposition, *Home Box Office* is a major victory for
the cable television industry and, indeed, vindicates arguments advanced by the Justice Department in this area for years. Whether
the decision will have much practical effect in making more competitive viewing options available to more of the public, however, is not
clear. The decision did eliminate regulatory constraints which to some
extent had inhibited the supply of feature films to the pay-cable industry, or had the effect of sharply reducing the time that pay operators
could exhibit a given movie, provided that they were able to get it.
Many of the movies that currently exist and are suitable for pay-cable
exhibition, however, are evidently under contract already to established interests. To be sure, future movie product may be more
readily available to pay-cable operators as a consequence of this decision. However, for the first year or so after a new movie is released,
it almost always is restricted by movie makers and distributors to the
theatrical circuit for economic, not regulatory, reasons. Thus, it may
be at least a year before any marked increase in the availability of
feature films to the pay-cable industry is plainly discernible.

Any such increase in available product to cable, moreover, is
partly dependent upon broadcast industry responses and purchases.
Currently, the television industry has asserted that it must buy long-
term exclusivity rights or "clearances" against pay-cable exhibition of
movies, for legitimate commercial reasons. To some extent, the business of the television industry, and especially that of the national
networks, consists not of "selling" audiences to major advertisers, but
rather of selling "audience futures." The structure of the industry,
coupled with the inelasticity of demand for national advertising
minutes, has resulted in a situation where advertisers book time on
networks far in advance. The television industry has argued that it
must be in a position to deliver on these audience futures contracts,
and that pay-cable competition, especially in major urban audience
areas, could adversely affect that ability on the part of the television
industry.

The question of what, if any, "clearance" broadcast television
may reasonably purchase against pay-cable television has been a

---

[8] Ibid., pp. 52-60.

matter of long-term dispute. Pay-cable operators have argued that no clearance or exclusivity is in any way justified, because of the disparate sizes of the industries involved. At the same time, pay-cable operators have argued somewhat incongruously that they should be permitted to purchase long-term exclusive rights against "free television" exhibition of movie products they may have bought.

In the *Home Box Office* ruling, the court took the extraordinary step of ordering the commission to examine the long-standing issue of what terms of exclusivity should be allowed, and to take final action on the matter within 180 days.[9] Whether the commission is capable of taking action in so short a time frame, or what kinds of rules it may adopt, is not clear. Whatever rules it might adopt, moreover, are unlikely to be retroactive in effect, and thus would not act to trim back the exclusivity provisions now contained in existing movie contracts.

There is an additional issue related to over-the-air pay television. Even "unleashed," pay-cable services will only be available as viewing options to those who first have access to a cable television system. In many of the major urban markets, where the preponderance of the population lives, the costs associated with deploying and maintaining cable television systems are staggeringly high. It is this matter, far more than the FCC's regulations, which accounts for the fact that few, if any, of the nation's cities have yet been wired. Although the ability to offer an expanded range of pay services may enhance the financial and commercial viability of urban cable television operations, the use of UHF television channels to provide over-the-air pay services is said by many to be a much more promising undertaking. Such UHF channels are presently available in many, if not most, of the likely target areas. The capital investment required to deploy such pay over-the-air television, or STV, operations is very much less than that which pay cable typically entails.

Optimism that as a consequence of the *Home Box Office* decision many more people will now have competitive viewing options available to them must be limited. This is because the D.C. Circuit affirmed the restrictions that the commission had adopted with regard to STV. Since the early 1960s, the FCC has imposed sharp restrictions on the use of television broadcast frequencies for pay-television operations. These rules are basically identical to those that were applied to pay-cable operations. The court essentially reasoned that, when use of an FCC-licensed radio frequency is involved in the distribution process, the commission may adopt almost any restrictions on the nature of the broadcasts that it chooses. As a result, the preponderance of the

[9] Ibid., pp. 9–10 and 105.

country will not in the foreseeable future have reasonable access to the kinds of uninterrupted movie exhibition services proponents maintain the public is ready and willing to buy.

Some Conclusions

We have endeavored to discuss many of the problems associated with achieving meaningful regulatory reform in cable television and to make available some of the DCRG's papers associated with the analysis of these problems so that policy makers in the future would not necessarily have to "reinvent the wheel." Cable television topics, however, should properly be looked at in the context of the problems associated with regulatory reform in general.

A majority of voting Americans evidently would agree that in the past twenty years, the level and intensity of government involvement in our national life has increased, is increasing, and ought to be diminished. The Carter administration, after all, was elected on a platform that included firm promises to cut back on wasteful and intrusive federal actions. Federal economic regulation, most would generally agree, is a prime example of such government waste and official intrusion.

It is axiomatic, of course, that one man's waste and intrusion, or "institutionalized cartel management," is another's "delicately balanced regulatory scheme." Donald I. Baker, former head of the Justice Department's Antitrust Division, once noted that

> many businessmen, especially in regulated industries, tend to look at competition and "deregulation" much the same way my children do green vegetables. Something that is no doubt very healthy, but for others. For themselves, they are much more interested in dessert.

There rarely is any significant consensus of opinion that supports specific, as opposed to generalized, deregulatory actions. The road through the Congress in this regard, moreover, is not paved with a history of significant accomplishments. Proponents for change in existing regulatory structures typically are required to confront subcommittee chairmen who are themselves the chief architects of the present regulatory system. Such audiences are not always, or even frequently, hospitable to recommendations for significant change, especially when advanced, as Paul W. MacAvoy has noted, mainly by "academics, nonelected bureaucrats who have 'never met a payroll,' or worse, economic theorists."

Regulatory changes in the mass media field are especially complicated and politically "sensitive" ventures. Broadcasting, and television broadcasting in particular, is widely regarded by both elected and nonelected politicians, as the most powerful of the celebrated Washington lobbies. Some evidence to the contrary notwithstanding, few politicians are willing gratuitously to assault the television establishment and thus, conceivably, jeopardize the most important of individual objectives—reelection, reappointment, or future employment by the industry.[10]

Virtually any legislative or regulatory proposal to alter or relax existing cable regulation automatically is perceived as being "antitelevision." This is true even though more than one-third of all cable television systems are actually owned by broadcasters, who conceivably would benefit from such relaxations. Whether or not the political influence of television exists (or, like "Hoover's Files," exists mainly in the popular imagination), the objective facts demonstrate that such proposals rarely enjoy a fair hearing on the merits.

Under such circumstances, the decisive role the courts have assumed in broadcast and cable regulation is not surprising. Since the *Carolene Products* case, at least, the judiciary has "subjected to more exacting judicial scrutiny" matters where ordinary political processes cannot reasonably be expected to implement necessary change.[11]

Predictions about the Future

Since completion of the Domestic Council's cable television program, the Communications Subcommittee of the House of Representatives has announced plans to overhaul much of the 1934 Communications Act and to consider specific legislation regarding cable television legislation. The subcommittee has stated that they will not consider piecemeal or interstitial legislative proposals prior to completion of a complete review, unless urgently required to do so. The subcommittee's comprehensive and ambitious effort is expected to take some time to complete, perhaps several years.

---

[10] See generally Thomas E. Patterson and Robert D. McClure, *The Unseeing Eye: The Myth of Television Power in National Elections* (New York: G. P. Putnam's Sons, 1976).

[11] United States v. Carolene Products Co., 304 U.S. 144, 152 and n. 4 (1938). See generally Brown v. Board of Education of Topeka, 347 U.S. 483 (1954) (desegregation); Baker v. Carr, 369 U.S. 186 (1962) (reapportionment); Phillips Petroleum Co. v. Wisconsin, 347 U.S. 672 (1954), and FPC v. Texaco, Inc., 417 U.S. 380 (1974) (energy regulation).

In the interim, it is reasonable to expect a gradual whittling down of existing FCC regulation of the industry, coupled with a commensurate increase in the regulation imposed by state, local, and municipal authorities. This in fact is already occurring. In an appeal brought by public utility commissioners, the FCC rules purporting to regulate various nonentertainment services which could be offered by cable systems were struck down on the grounds that the commission had no statutory authority to regulate such services and that the assertion of authority unlawfully restricted legitimate state utility regulation.[12]

There is no question but that cable television as an industry will continue to grow. Pay-cable services are also likely to expand. Indeed, even subject to what were maintained to be onerous FCC regulations, and an imposed "starvation diet," the pay-cable industry roughly doubled in size during the last seven months of 1976, and the number of cable subscribers also subscribing to pay services was growing by some 7,000 new customers a week. In this respect, a spokesman for CBS has suggested that the cable television industry may be at some pains in the future to explain away a failure to expand geometrically once the regulatory constraints "holding cable back" are lifted.

As the cable television industry grows and matures, regulation of the industry will become a more difficult and complicated undertaking. Certainly, as more and more citizens have CATV services available to them, the probability that Congress would act to withdraw or constrain those services greatly diminishes. As one commentator noted, the public, "once having tasted the lollipop, will not lightly suffer Congress taking it away from them." What is much more probable, of course, is that the cable television industry will come to appreciate the clear and demonstrable benefits a regulatory thicket can afford, especially if that thicket is built on and operated like traditional models. The benefits which traditional regulation can offer to regulated industries are well known and in some quarters notorious. Most likely, the cable television industry will come to the recognition of this fact, if they have not done so already. The public policy concern at such future time, then, will be with how to forestall the industry, imposing its own version of a regulatory briar patch to protect cable television from future, potential competitors.

---

[12] National Association of Regulatory Utility Commissioners v. FCC, 533 F.2d 601, 606-07 (D.C. Cir. 1976).

APPENDIX A

Letter of the Department of Justice on Cable Television Legislation

Department of Justice
Washington, D.C.
June 17, 1975

Honorable James T. Lynn
Director
Office of Management and Budget
Washington, D.C. 20503

Dear Mr. Lynn:

This is in response to your request for the views of the Department of Justice concerning the Office of Telecommunications Policy's most recent draft bill "To amend the Communications Act of 1934 to create a national policy respecting cable communications." This complex and lengthy proposed legislation has been carefully reviewed. For clarity's sake, our report here is in three parts.

Part 1 endeavors to review the various aspects of this draft legislation in light of the specific comments and suggestions the Department submitted with respect to OTP's earlier cable bill. Those comments and suggestions are set forth in our August 5, 1974, letter to your predecessor, and are for the most part relevant here, inasmuch as this draft bill differs from its predecessor chiefly in format. There we limited our remarks to specific points we will call the antisiphoning issue, the cross-ownership issue, the interconnection issue, and the separations issue.

This letter over the signature of Acting Assistant Attorney General A. Mitchell McConnell, Jr., is reproduced here in its original form with only a few minor editorial modifications.

Part 2 discusses aspects of this proposed legislation from the standpoint of constitutional sufficiency, and will touch as well upon the serious impact this bill could have on the Federal judiciary. Time did not permit us to prepare comments on these issues in our earlier report to your predecessor.

Part 3 discusses several additional policy questions implicit in this proposed legislation which we did not have time to consider in our report on the original OTP draft bill. Members of your staff have suggested that in this report we endeavor to address those policy issues and questions because of the potential significance of this legislative initiative. Having participated in 1971–72 with OTP in formulating some of the basic policy outlines of the 1974 Cabinet Committee report, we are somewhat reluctant to undertake at this time any wholesale review of the assumptions upon which this bill has been based. Accordingly, we have tried to limit our comments in this regard to the matters we feel are most important. Overall, we believe it is necessary to bear in mind that in the four years since the policy outlines of that report were sketched, general economic and cable industry circumstances, as well as the prevailing regulatory environment, have very markedly changed.

Among other things, the growth of cable television, then seen as likely to be rapid, has in fact greatly slowed, and the prospect of its near term "explosion," has greatly diminished. Currently, we are informed, cable industry growth is almost entirely internal—i.e., adding subscribers to existing systems—and the expansion of this capital intensive industry in terms of new systems is at a virtual standstill. Perhaps, the principal reason for the slowdown in cable's growth has been the development or extension of policies by the FCC which seek to restrict the role of cable to that of a *supplement* to the over-the-air broadcasting industry, rather than allowing cable a reasonably free opportunity to develop on the merit of its service offerings. Since 1971, the FCC has adopted increasingly pervasive and restrictive cable regulations. Indeed, it is paradoxical to observe that during this same four-year time frame the FCC has initiated and vigorously pursued a policy of relaxing broadcast regulation. *See, e.g., Re-regulation of Radio and Television Broadcasting*, 37 Fed. Reg. 23723 (1972), 37 Fed. Reg. 11537 (1972), FCC Mimeos 74-261, 74-653 (1974). The Commission can thus be seen as working to reduce regulatory burdens regarding the industries its statutory mandate specifically encompasses, while on the other hand endeavoring to use its slender ancillary jurisdictional thread to weave an increasingly constraining web about cable. *Cf. National Cable Telev. Assoc., Inc.*

v. *United States*, 415 U.S. 336 (1974); *TelePrompTer Corp.* v. *Columbia Broadcasting System, Inc.*, 415 U.S. 394 (1974). In view of these objective facts and the Administration's ongoing efforts to achieve meaningful regulatory reform generally, we believe that if any effort is to be made by the Administration to introduce cable legislation at this time, such legislation should more clearly limit the FCC's regulatory authority to restrict the development of cable than would the bill proposed by OTP.

We will to some degree expand upon these general comments in subsequent parts of this report, and will detail what we see as their possible implications.

The Department's Earlier Comments

In our August 5, 1974, report to your predecessor regarding this proposed legislation, we commented on four basic issues. First, we objected to the provisions which proposed, in essence, to codify the FCC's highly restrictive antisiphoning rules, those rules which limit the types of programming that may be shown on cable for a charge separate from the monthly subscription charge. We stated that such restrictions should not be imposed absent some clarification of the exact public interest sought to be protected, and a clear showing that no less anticompetitive alternatives would be adequate to safeguard that public interest.

Second, we objected strongly to provisions of the earlier draft bill which would have prohibited the FCC from prohibiting or limiting cross and multiple ownership of cable systems and other media, particularly the language which could have been construed as creating a new (and, in our view, wholly unwarranted) antitrust exemption. Congress has specifically recognized the importance of preserving competition in broadcasting, 47 U.S.C. §313, and the Supreme Court has repeatedly held that the preservation of competition in broadcasting is a component of the statutory public interest test which the Commission must consider. *See, e.g., National Broadcasting Co.* v. *United States*, 319 U.S. 190, 222–224 (1943), and *United States* v. *RCA*, 358 U.S. 334, 351 (1959). When considering cable television, moreover, it is important not only to focus upon the loss of actual competition which such media cross-ownership can entail, but also to bear in mind the loss of economic incentives fully to exploit the potential of cable. Put simply, firms having vested interests in existing media technologies may not press, or press as hard, for the emergence of cable services that may be competitive with their commonly owned, established press or broadcast business. To assure maximum entrepreneurial incentives therefore, we firmly believe that the ownership of cable television should be separate at this time as a general matter from significant ownership of other local media. Such prohibitions would not prevent broadcasters and newspaper publishers from investing capital in the cable industry. They would be free to invest in cable operations in every market in which they did not own a broadcast license or publish a paper. Indeed, it is our understanding that a number have invested in cable systems in distant markets. If, in the future, the ownership of cable systems should become substantially separated from the programming on such systems, the need for prohibitions on common ownership of cable and other co-located media, there is no present reason to prohibit the FCC from adopting rules designed to preserve local media competition if the Commission determines that such rules are necessary.

Third, we suggested a recasting of the bill's language with respect to terminal interconnection. We recommended that the interconnection of equipment be permitted unless the FCC found that specific equipment was in fact technically incompatible with a cable system. As the bill had been drafted, FCC equipment type approval would have been required as a prerequisite to its use in conjunction with a cable system.

Fourth, we suggested that in view of the paramount importance of the separations policy (i.e., the goal of separating ownership of a cable system from programming control) to this proposed legislation, language should be added that would indicate clearly when, and in what fashion, implementation of this policy should actually occur. We pointed out the risks of relying upon broad administrative discretion to determine when some imprecisely defined "triggering" development has occurred.

Of the four major issues addressed by the Department in its comments on the original OTP draft bill, only the provisions relating to the interconnection of terminal equipment have been changed to satisfy the Department's concern.

While the specific language which seemed to codify the FCC's restriction on the types of programs that may be shown on cable for a charge separate from the monthly subscription fee ("pay cable") has been deleted from the revised OTP draft, the Department's concerns have not been addressed. We had urged the view that any cable legislation should articulate a clear Congressional policy limiting the power of the FCC to restrict pay cable programming to those restrictions needed to preserve the minimum amount of over-the-air broadcasting necessary to serve the public interest in the particular locality. Rather than provide such clear policy guidance, the revised OTP draft ducks the whole issue. The OTP section-by-section analysis of its revised bill candidly acknowledges that the courts will have to decide what, if any, limits the general language of sections 103(d) and 401(a) of its revised draft bill would have upon the FCC's authority to restrict pay cable to protect over-the-air broadcasting from program siphoning. We believe that deliberately leaving the resolution of such an important issue as what is Congress' policy on pay cable to years of litigation in the courts is inconsistent with this Administration's general policy of reducing the delays and obstacles to private decision making imposed by governmental action.

This revised draft bill does make some changes in the earlier language with respect to the issues of cross-ownership and multiple ownership of cable systems. Unfortunately, the revisions, as drafted, can be viewed as exacerbating some of the competitive concerns previously outlined. Section 204(d), for example, would, if enacted, prohibit the FCC from barring the ownership or control of cable systems by national television networks or co-located television stations, provided the owner-operator offers no more than what could be called a "basic cable services package"—essentially, retransmitting local signals (or the signals the FCC may prescribe), providing an access channel, and originating programming on a single channel. Thus, new section 204(d) can be viewed as going beyond former section 406(f), which would have merely grandfathered existing TV station-cable system cross-ownerships.[1] Section 401(d) of this draft bill relates to cable system ownership by publishing interests, for example, co-located daily newspapers. If enacted, it would prohibit any Federal executive agency (including the FCC and possibly even the Department of Justice) from adopting so much as a policy against such cross-ownership, provided such systems offered only the services specified in section 303(g)(2). That subsection, in turn, "requires"

---

[1] Existing FCC rules currently prohibit national television networks from cable system ownership. 47 C.F.R., section 76.50(a)(1972). See Columbia Pictures Industries, Inc., 30 FCC 2d 9 (1971). These same rules also require dissolution of TV-cable combinations in the same community. Out of the Nation's 3,000 plus cable systems, 200 are apparently owned by a co-located TV station; however some of these systems are in major markets. The Commission has recently proposed an amendment to the rules which would prohibit the common ownership of co-located television stations and cable systems in the future but would grandfather current cross-ownership situations except where the commonly owned television station places the only city-grade signal over the cable system's community.

such systems to offer only a "modified" basic cable services package—transmitting local signals, providing an access channel, but originating "*no less than* [*sic*] two channels" of programming. (Emphasis supplied.)

For the reasons indicated above, the Department does not believe that it is either wise or necessary to prohibit the FCC from adopting rules designed to prohibit concentration of control with respect to the various communications media. Regulation designed to preserve competition must be distinguished from regulation designed or having the effect of restricting competition. This Administration has been urging regulatory agencies to preserve and promote competition to the maximum extent consistent with regulatory goals. *See Economic Report of the President*, pp. 147–159 (1975). Experience under the Communications Act, to date, has not revealed any examples of situations in which the FCC has unwisely limited common ownership of media at the expense of other important public interests. Consequently, there is no need to prohibit the FCC from taking such action if it determines that such measures are in the public interest. Therefore, we urge the deletion of the language in section 204(d) of the revised OTP draft bill which would prohibit the FCC from limiting the ownership of cable systems by national television networks and co-located television stations.

The Department's objections to section 401(d) of the revised OTP draft bill are even stronger than our just noted objections to section 204(d). Section 401(d) prohibits under prescribed circumstances "any executive agency of the United States, including the Commission" from adopting any rule, regulation or policy prohibiting or limiting the ownership of cable systems by publishing interests or owners of other cable systems. Here too, we believe that there is no need to prohibit the FCC from taking such action as it deems necessary to protect the public interest in competition between media. However, the language of section 401(d) is such that it could be construed as prohibiting the Department of Justice from bringing suit, thereby creating an antitrust exemption.

The Department objected to analogous (though more sweeping) language in the earlier draft, pointing to the potential difficulties of interpretation such language could create. *Cf. Ricci* v. *Chicago Mercantile Exchange*, 409 U.S. 289, 303 (1973); *Carnation Co.* v. *Pacific Westbound Conf.*, 383 U.S. 213, 218 (1966). We also pointed out that the Department had filed an antitrust action against the proposed merger of two of the major multi-system cable operators (*United States* v. *Cox Cable Commun.*, Civ. No. 17573 (N.D. Ga. 1972)) and could perceive of no public purpose to be achieved by foreclosing judicial consideration of such issues. The same objections seem relevant to this revised language. Accordingly, we recommend that section 401(d) be deleted in its entirety. At the least, if section 401(d) is retained, there should be added to this draft bill language providing substantially as follows:

> Sec. 402. Nothing herein shall be deemed to affect or in any way to derogate from the responsibilities of the Department of Justice, the Federal Trade Commission, or the Federal Communications Commission to enforce the antitrust laws or any other laws of the United States.

Unfortunately the revised OTP draft bill also fails to satisfy the Department's previously expressed concern over the lack of clear guidance provided in the language which was designed to implement the policy of separating ownership of a cable system from control of the programming provided over the system. The OTP revised bill would direct the cable licensing authority to require a cable system to have a specified minimum number of channels, section 303(g)(2). Any channel capacity in excess of that reserved for the system operator under section 303(g)(2) would have to be made available for lease by independent programmers, section 303(f). Section 303(g)(1) then provides that the cable licensing authority shall require that the proportion of the entire system made available for lease to independent programmers be increased by terms set out in the license which "shall be reasonably related to the use of, and unfulfilled demand for, such capacity." In our opinion, this general directive to the local or state cable licensing authorities leaves to them the degree to which a separations policy should be pursued without any clear Congressional guidance.

Judicial Impact and Related Issues

In our earlier comments, the Department reserved comment with respect to certain aspects of this bill, including its proposed handling of certain Federal-State delegations, and its impact on general Federal judicial administration.

Of these issues, section 502 of the draft bill is perhaps most troublesome. That section, if enacted, would essentially waive diversity-of-citizenship and amount in controversy prerequisites to initiating actions to enforce this bill in Federal district courts. No "judicial impact" information regarding this provision has been pro-

105

vided by OTP, although the FCC in its earlier comments asserted that this section could result in "further glutting" the Federal district courts. (*See* FCC sectional analysis at p. 13.)

The Department, as you know, has for some time now expressed serious concerns regarding the very heavy caseloads currently placed upon the Federal district courts, and Congress' reluctance promptly to authorize an increase in the number of Federal judges. Currently, over 100,000 civil and some 40,000 criminal cases are commenced annually in the district courts. *See, e.g., 1974 Report of the Director of the Administrative Office of the U.S. Courts,* at tables 18 (p. ix-26), 47 (p. ix-65). Yet the number of judges has been held to approximately 450, including both active and senior judges. Moreover, recent Congressional enactments, such as the expediting provisions in the new Freedom of Information Act Amendments (P.L. 93-502) are likely to exacerbate the continuing problem of overburdened court calendars.

Under these circumstances, therefore, the Department believes it would be most unwise to expand Federal court jurisdiction without a demonstration of a serious need to do so and a careful effort to weigh the additional burden.

In addition, we have considered two constitutional problems which may be raised by this bill. The first is whether despite the conceded Federal regulatory power in the area of CATV the particular scheme employed can be said to violate constitutional principles of federalism by requiring States to act at the direction of the Federal Government; the second is whether the jurisdiction conferred on Federal courts by sections 501 and 502 of the bill is within the limits of Federal court jurisdiction provided by Article III of the Constitution.

(A) The first potential constitutional question arises from the fact that the draft bill establishes a dual regulatory system for cable communications and defines the term "cable licensing authority" as including a State itself. Section 202 provides that no person may operate a cable system initially licensed or relicensed after the effective date of the act without a certificate of compliance issued by the FCC. Section 203(a) states that the FCC first must determine whether the applicant for a certificate holds a valid license to construct and operate a cable system and such licenses would be issued exclusively at the State or local level of government by a "cable licensing authority," section 301. "Cable Licensing authority" is defined as "any state, county, municipality, or any political subdivision thereof, . . . that is empowered by law to authorize by license, . . . the construction and operation of a cable system within the jurisdiction of such agency," section 105(f). A "cable licensing authority" must adopt a system of licensing which follows the standards and requirements set forth in the bill, section 303.

As a matter of inherent constitutional powers, a State may be regarded as "empowered by law" to issue cable franchises. Consequently, on one plausible interpretation of the draft definition of "cable licensing authority" a State would become a cable licensing authority which is required by section 303 to adopt a system of licensing in conformity with Federal requirements regardless of whether it had enacted a system of cable regulation or had any desire to carry out such regulation. On that interpretation, the Federal Government would in effect be directing the State that it must legislate, denying the legislators the ability to choose between legislating or not and leaving them but ministerial agents of the Federal Government. Such a scheme certainly would do violence to the principles of federalism implicit in the Constitution.

Just as is the case with regard to Federal taxation of State activities, there is a constitutional limit to the Federal Government's ability to impinge on a State's sovereignty even in an area involving one of the specified powers of Congress. As Chief Justice Marshall said in *McCulloch* v. *Maryland*, 4 Wheat. 316, 421 (1819):

> Let the end be legitimate, let it be within the scope of the constitution, and all means which are appropriate, which are plainly adapted to that end, *which are not prohibited, but consistent with the letter and spirit of the constitution*, are constitutional. (Emphasis added.)

See also, New York v. *United States*, 326 U.S. 572, 586 (1946) (Stone, Ch. J.).

Since we are certain that no such result was intended by OTP, we believe that the problem could be dealt with adequately by changing subsection 105(f) to read:

> "Cable licensing authority" means any agency, commission, board, or authority of any State, county, municipality, or political subdivision thereof, that is empowered by law to authorize by license, franchise, permit or other instrument of authority, the construction and operation of a cable system within the jurisdiction of such agency.

(B) The second constitutional question raised by the bill is whether Congress may, consistent with Article III of the Constitution, confer jurisdiction in sections 501 and 502 on Federal courts to decide questions of State law absent diversity of citizenship. Sections 501 and 502 confer jurisdiction on the district courts of the United States, without regard to the citizenship of the parties or the amount in controversy, over actions challenging "any act, practice, or omission of a cable licensing authority or cable operator . . . on the ground that it does not comply with the provisions of this act or the provisions of a statute, ordinance or law of a state, or political subdivision thereof, intended to implement or apply the provisions of this act." Unless the jurisdiction granted a Federal court by statute can be found to be within Article III, that statute would be unconstitutional. *Marbury* v. *Madison*, 1 Cr. 137 (1803). The jurisdictional grant of sections 501 and 502, however, should not be reviewed in the abstract, and as a practical matter a purely State law question by itself is unlikely ever to arise under the bill.

Almost all cases challenging acts or omissions under State cable law or regulations would presumably claim, at least among other things, either that the State laws or regulations were not consistent with the Federal law or that the State licensing authority or cable operator had acted in a manner inconsistent with Federal law. That is, even claims challenging acts pursuant to State regulations will most likely raise in the pleadings questions of interpretation of the Federal law. This at least has been the experience in welfare law, which in terms of the regulatory framework is not dissimilar from the proposed bill. Such claims, even though challenging acts under State regulations, essentially raise questions regarding the Federal law as applied and as such qualify on traditional grounds for Federal question jurisdiction. *See, e.g., Gully* v. *First National Bank*, 299 U.S. 109, 112–13 (1936) (Cardozo, J.). Even where a particular claim is purely one of State law, e.g., whether there is substantial evidence to support a finding, it will almost invariably be pendent to a claim that would raise a Federal question. Even after the court determines the Federal question, it may go on to decide the purely State question on the basis of pendent jurisdiction. *See, e.g., Rosado* v. *Wyman*, 397 U.S. 397 (1970); *Gem Corrugated Box Corp.* v. *National Kraft Container Corp.*, 427 F.2d 499 (2d Cir. 1970). *See also UMW* v. *Gibbs*, 383 U.S. 715 (1966).

Finally, even if a court were presented with a case purely of State law which did not implicate the Federal law in any way, the court would probably not find the jurisdictional provision in the proposed bill unconstitutional. Rather it would probably interpret it in such a way as to keep it constitutional. *See, e.g., Regional Rail Reorganization Act Cases;* ... U.S. ..., ... (1974), 43 USLW 4031, 4041 (S. Ct. December 16, 1974). Thus, it might interpret the language of section 501 as creating a cause of action in Federal courts where an act or omission under State law is challenged in such a way as to raise a question of interpretation of the act itself, even if only as applied; this would create a traditional Federal question.

Evaluation of Additional Policy Questions Implicit in the Revised Draft Bill

Your staff has requested that we undertake to evaluate any policy questions implicit in the bill which we did not discuss in our report on the original OTP draft bill. Before doing so, however, it seems appropriate to review again briefly the basic policy recommendations contained in the January 1974 "Report to the President of the Cabinet Committee on Cable Communications," prepared by OTP. That report, as you will recall, recommended adoption of essentially two basic policies with respect to Government regulation of cable television.

First the report urged adoption of a so-called "separations policy," sometimes also referred to as "common carrier status for cable." By separating ownership and control of cable programming from ownership and control of the cable system itself, the authors of this report reasoned any potential for monopolistic domination over the content of cable communications would be diminished. *See generally,* Kestenbaum, *Cable Television as a Common Carrier* (paper presented at the OTP Conference on Communications Policy Research, November 18, 1972).

A second important feature of the Cabinet Committee report is the recommendation that there be adopted an institutional framework that would apportion any necessary government regulation among Federal, State and local authorities. The report saw this "cooperative Federalism" approach as achieving among other things an important objective—namely, reversing what it viewed as an undesirable trend toward increasingly obtrusive Federal involvement in regulating electronic media. As a matter of regulatory policy, moreover, the report, pp. 12-13, recognized the need for developing a regulatory scheme characterized by reasoned specificity, rather than as in the past, relying upon undirected and largely unfettered administrative discretion.

The Department appreciates the merit of a separations policy in cable once the industry has progressed to the point of financial viability. The most difficult question, in our opinion, is the manner and time span over which such a policy is implemented. If a separations policy is implemented too fast, it may actually impede, if not defeat, the ability of cable to provide greater diversity in our communications system. In the current market and regulatory environment, the cable operator, and not third parties, would appear to have the clearest incentive to supplement the system's basic offerings with additional programming. Proscribing this activity, therefore, could adversely affect both the current operations and growth potential of cable. Even though this draft legislation does not actually propose to implement the full separations policy at this time, it does impose certain conditions upon system operator programming which could deter such activity. While the system operator may originate programming beyond his basic service complement, section 303(d) requires that new systems provide capacity for lease by independent programmers equivalent to the amount used by the operator. Under present economic and regulatory conditions, there is little channel leasing being done by independent programmers. The FCC rules currently do not prohibit the leasing of channels by independent channel programmers, yet fewer than 100 cable channels are presently so leased nationwide.[2] Thus, under the OTP bill, cable operators who desire to provide a significant amount of programming themselves will have to incur the expense of building a system with channel capacity which may prove to be considerably in excess of demonstrated market demand.

This apparent dearth of independent channel programmers, moreover, may have serious potential implications when considered in light of the local media cross-ownership this proposed legislation would allow. For, absent the emergence of such a class of vigorous entrepreneurs, the result may be the achievement in many major metropolitan areas of precisely the very problem that the Cabinet Committee sought basically to avoid. For illustration's sake, assume that the cross-ownership provisions of this bill were enacted, and that a newspaper-TV-radio combination serving a given community acquired control of a co-located cable television system—a system which this bill would essentially declare to be the "natural monopoly" system for the community (*see* section 102(d)). Unless, however, a class of entrepreneurs eager to lease and program cable channels should be present, the plain result would be to produce a single firm controlling nearly all channels of communications within the community—exactly what the Cabinet Committee sought to avoid. That is, the channel capacity would be there for lease, but there would not be anyone willing to lease it.

We realize, of course, that a strong channel leasing programming industry may emerge. One reason why it has not developed to date may well be the FCC's so-called "pay cable" rules. Those rules, as

---

[2] Statement of FCC Cable Bureau representatives, at the Commerce-Justice-FCC-OTP-OMB Conference on Cable Legislation, Thursday, August 15, 1974.

the Department has repeatedly pointed out, reflect a protectionist policy determination that practically all currently popular programming be reserved to over-the-air broadcasting. Given such inflexible regulatory constraints, it is certainly not surprising that a significant channel leasing programming industry has not yet developed. As the Commission has recognized, such an industry is to a great extent dependent upon the entrepreneur's opportunity to program for pay, since it is unlikely that it will be afforded any share of cable subscriber revenues or be able to sell sufficient sustaining advertising.

We have already commented upon the fact that several important provisions of the proposed bill lack the clear policy guidance needed to settle controversial issues. Moreover, we believe that this draft bill does not, contrary to this Administration's general goal, significantly reduce the amount of governmental regulation. For, under Title II of the bill, many of the authorities presently asserted by the Commission would only be affirmed and ratified. The FCC would be specifically empowered, for instance, to determine exactly which signals a cable should be permitted to retransmit (section 204(a)), an authority which would afford the FCC basis for continuing its present welter of rules on "mandatory carriage" and "network exclusivity." Indeed, the bill would even seem to permit continuation of the so-called "distant-signal importation" rules—described by one expert as evincing a "policy which appears to be that the only way to be permitted to import distant signals is to demonstrate that no one will watch them!"[3] Virtually every other important facet of the FCC's present cable television regulatory regime (excepting the existing rules regarding cross-ownership), moreover, appears included in this Title. Indeed, all powers not explicitly provided for, could be asserted by the FCC, under the proposed "umbrella" power to grant cable systems "certificates of compliance," for it cannot lightly be assumed that the power to grant would exclude the power to condition a grant, or indeed, to modify or revoke it. (See generally, Charles River Bridge v. Warren Bridge, 36 U.S. (11 Pet.) 420, 547, 553 (1837); National Broadcasting Co. v. United States, 319 U.S. 190, 215–16 (1943).

Other provisions of this proposed legislation can be viewed as doing little more than ratifying authorities now exercised by many local cable television franchising authorities. Section 401(f), for example, would explicitly authorize local authorities to continue to prescribe "reasonable" rates for various cable services without regard

---

[3] S. M. Besen, *The Economics of the Cable Television "Consensus"* (Paper presented at the OTP Conference on Communications Policy Research, November 18, 1972), p. 13.

to the amount of competition facing the system operator. The same section, moreover, lays a foundation for the eventual regulation of channel leasing charges, a proposal which seems somewhat inconsistent with language in the Cabinet Committee report itself. (*See* report at 42-44).

It may well be, as OTP has suggested, that this proposed legislation would be desirable for no other reason than it might apply a partial restraint upon the steady increase in the detail and pervasiveness of Federal regulation of cable television. Such a halt or moratorium on further regulation would, of course, be an objective the Department would support. It is not clear, however, that this proposed legislation would in fact set any such firm outer bounds. Rather, in some ways it can be viewed as we have suggested as proposing little more than ratification of the regulatory status quo as it pertains to cable.

Conclusion

In this report, as your staff has suggested, the Department has discussed some of the serious policy questions that are raised by this draft bill. Among these questions is the overall desirability of proposing legislation which leaves most of the restrictive existing regulation of cable in place and fails to articulate clear Congressional guidance on certain important issues. For the reasons set out above, the Department does not believe that the Administration should introduce or support the OTP cable bill as presently drafted.

Sincerely,
A. MITCHELL McCONNELL, JR.
Acting Assistant Attorney General

APPENDIX B

Draft Cable Television Legislation Proposed by the Office of Telecommunications Policy, August 1975

A BILL

To amend the Communications Act of 1934 to create a national policy respecting cable communications.

Be it enacted by the Senate and House of Representatives of the United States of America in Congress assembled:

Title I. Short Title, Findings, Declaration and Application of Policy, and Definitions

Short Title

Section 101. This Act may be cited as the "Cable Communications Act of 1975."

Findings

Sec. 102. The Congress hereby finds:

(a) that cable systems are engaged in interstate commerce through the origination, transmission, distribution, and dissemination of television, radio, and other electromagnetic signals, and have the technological capacity to provide a multiplicity of varied communications services;

(b) that the expansion, development, and regulation of cable communications, while a matter of importance to non-Federal governments, is also of appropriate and important concern to the Federal Government;

This draft bill is reproduced here in the form and editorial style in which it was prepared by the Office of Telecommunications Policy.

(c) that the application of policies intended for broadcast communications is inappropriate for cable communications in that cable technology eliminates the channel scarcity found in television broadcasting, and thus facilitates the provision of programming and other communications services not otherwise available over broadcast facilities;

(d) that, given technical and economic considerations, cable systems are likely to evolve as natural monopolies within their service areas;

(e) that it is necessary and appropriate for the Federal Government to establish and support a national cable policy to provide fair opportunity for the evolution of cable as a medium in its own right, open to all and free from both excessive concentrations of private power and government restrictions that would deny the public the full potential of cable services; and,

(f) that a national and uniform policy is needed for cable to prevent the emergence of conflicting or duplicatory regulation by various governmental authorities.

Declaration of Purposes and Policy

Sec. 103. The Congress accordingly declares that the purposes of this Act are to:

(a) initiate an evolutionary plan, pursuant to its powers to regulate interstate commerce, that will result in the adoption of a comprehensive national policy to allow the growth and development of cable communications which will be responsive to, and serve the needs and interests of, the public;

(b) create a regulatory framework, as the first step in such an evolutionary plan, which would apportion the authority to regulate cable systems between the Federal Government and a non-Federal level of government, and would provide uniform standards and guidelines for the exercise of such regulatory authority by the several states or their political subdivisions;

(c) assure that cable develops as a communications medium open to all, free of both excessive concentrations of private power and undue government regulation and control that would inhibit the communication of information and ideas, or otherwise deny the public the full benefit of the services to be provided or offered over cable systems;

(d) assure that cable is provided fair opportunity to develop as a medium of communication in its own right, allowed to compete freely in the marketplace with other communications media, and is not, in relation to such other media, limited by regulation to an auxiliary or supplementary role in the provision of communications services to the public;

(e) establish as the goal of the evolutionary plan initiated by this Act the eventual separation of control of cable system from control of the content of cable channels, so that such content may be insulated from any local monopoly power of a cable system, as well as from the government regulatory power that otherwise would be necessary to prevent abuses of such monopoly power;

(f) assure that cable is regulated at the Federal and the non-Federal levels of government in a manner designed to achieve the National policy goal of eventually separating control of cable systems from control of the content of cable channels.

Application of Policy

Sec. 104. The provisions of this Act shall apply as follows:

(a) Any cable system constructed on or after the effective date of this Act shall be subject to the provisions of this Act and such rules, regulations, or orders, as may be adopted pursuant thereto.

(b) Any cable system constructed prior to the effective date of this Act shall be subject to its provisions, unless otherwise provided in the Act, and shall be subject to such rules, regulations, or orders as may be adopted pursuant thereto when its license period ends, or two years from the effective date of this Act, whichever shall occur first.

(c) A state, or any duly empowered political subdivision or agency, board, commission or authority thereof, may adopt or continue in force any law, rule, regulation, order, or standard affecting cable systems, *provided* that such law, rule, regulation, or order or standard is consistent with the exclusive grants of authority under Title II and Title III of this Act; is not prohibited by Title IV of this Act; and does not otherwise create an undue burden on the interstate commerce in cable communications.

Definitions

Sec. 105. For the purposes of this Act,

(a) "Cable system" means a facility or combination of facilities under the ownership or control of a single person or entity and authorized to serve a particular geographic area or location, which consists of a primary control center used to receive and retransmit, store, process, and forward, or, to originate radio, television, and other elec-

115

tromagnetic signals; and transmission facilities, with multi-channel capacity, used to distribute or otherwise disseminate such signals over one or more coaxial cables or other closed or shielded transmission media from the primary control center to a point of reception at the premises of a cable subscriber; *provided* that such term shall not be understood or construed to include such a facility or combination of facilities that:

(1) serves fewer than 500 subscribers; or

(2) serves only to retransmit the signals of radio and television broadcast stations, defined as local stations by the Commission.

(b) "Cable channel" or "channel" means that portion of the electromagnetic frequency spectrum used in a cable system for the propagation of a radio, television, or other electromagnetic signal.

(c) "Multi-channel capacity" means the capacity of a cable system to transmit simultaneously the equivalent of five or more television signals.

(d) "Closed transmission media" means media having the capacity to transmit simultaneously electromagnetic signals over a common transmission path such as a coaxial cable, optical fiber, wire, waveguide, or other signal conductor or device.

(e) "Cable operator" or "cable system operator" means any person or entity, or an agent or employee thereof, that operates a cable system, or that directly or indirectly owns an interest in any cable system; or that otherwise controls or is responsible for, through any arrangement, the management and operation of such cable system.

(f) "Cable licensing authority" means any agency, commission, board, or authority of any state, county, municipality, or political subdivision that is empowered by law to authorize by license, franchise, permit, or other instrument of authority, the construction and operation of a cable system within the jurisdiction of such agency.

(g) "Person" means an individual, partnership, association, joint stock company, trust, or corporation.

(h) "Interconnection facilities" means microwave equipment, boosters, translators, repeaters, communications space satellites, or other apparatus or equipment used for the relay or transmission and distribution of television, radio, or other electromagnetic signals to a cable system.

(i) "Cable subscriber" means any person who, for payment of a consideration, receives radio, television, or other electromagnetic signals distributed or disseminated by a cable operator or a channel programmer over a cable system.

116

(j) "Channel programmer means any person who leases, rents, or is otherwise authorized to use the facilities of a cable system for the origination of programming or other communications services over a cable channel, except the use of a channel by a cable subscriber to transmit an electromagnetic signal. Such term shall include a cable system operator to the extent that such operator, or person or entity under common ownership or control with such operator, is engaged in program origination.

(k) "Origination" or "program origination" means the use of a cable channel by a channel programmer for the distribution or dissemination of any program, including a pay cable origination, or other communications service, except the retransmission of the signals of a radio or television broadcast station by the cable operator or any person or entity under common ownership or control with such operator.

(l) "Pay cable origination" means the use of a cable channel by a channel programmer for the distribution or dissemination of any television program or other communications service, to a cable subscriber for a fee that is separate from and in addition to any fee charged such subscriber for the installation, connection, or maintenance of a cable system or for the distribution or dissemination of the retransmitted signals of broadcast television or radio stations or other program origination.

(m) "Cable license" means the license, franchise, permit, or other authorization issued to a cable system by a licensing authority.

(n) The term "Commission" means the Federal Communications Commission.

(o) The term "State" includes the District of Columbia, the Commonwealth of Puerto Rico, and territories and possessions of the United States.

Title II. Authority, Responsibilities, and Functions of the Federal Communications Commission

Authority

Sec. 201. (a) The Federal Communications Commission shall have exclusive authority to execute and enforce the provisions of this Title.

(b) The Commission may require any cable system constructed prior to the effective date of this Act to conform to the provisions of this Title and any rule, regulation, or order of the Commission adopted pursuant thereto.

(c) The Commission may continue or adopt such other rules, regulations, or orders respecting cable system facilities excluded from the provisions of this Act pursuant to subsections 105(a)(1) and (2) as may be appropriate and consistent with its responsibilities and pursuant to law.

Certificate of Compliance

Sec. 202. No person shall operate a cable system constructed after the effective date of this Act unless such person is issued a certificate of compliance by the Commission. The Commission shall grant certificates of compliance upon application therefor and upon its determination that such application is consistent with the provisions of this Title and such rules and regulations as the Commission may adopt pursuant thereto, and that the applicant for such certificate of compliance holds a license to construct and operate a cable system issued by a cable licensing authority.

Responsibilities and Functions

Sec. 203. In order to carry out the purposes of this Act, the Federal Communications Commission shall adopt or continue in force appropriate rules, regulations, and orders that:

(a) establish minimum technical standards and assure conformance to such standards by cable system operators and equipment manufacturers as may be necessary to promote the compatibility and interoperability of cable systems, the compatibility of the receivers or other terminal equipment connected to such systems by cable subscribers, and to prevent harmful interference to radio communications;

(b) assure that cable subscribers may determine in advance of reception the nature of programming originations and may, at their request, be provided appropriate means at reasonable cost to such subscriber to preclude, avoid, or control intelligible reception of program originations such subscribers do not wish to receive.

Sec. 204. The Federal Communications Commission may:

(a) require by rule, regulation, or order, notwithstanding the provisions of Section 152(b) of Title 47 of the United States Code, the use of any poles, ducts, conduits, or other such rights-of-way owned or controlled by any public utility, including telephone common carriers or other persons engaged in the provision of water, gas, or electrical energy services to the public, by a cable operator for the purpose of constructing, operating, or maintaining the transmission facilities of a cable system, pursuant to such terms and conditions and for such compensation as the parties affected may agree upon, or, in the event of a failure to so agree, as the Commission may fix as just and reasonable for the use so required, to be ascertained on the principle controlling compensation in condemnation proceedings; *provided* that the Commission finds that the business, service, or performance of such public utility is not substantially impaired by such use, and that it is otherwise practicable and consistent with the public interest; and *further provided* that any such public utility whose poles, ducts, conduits, or rights-of-way have been used by a cable system operator pursuant to a rule, regulation, or order of the Commission issued under this section, shall be entitled to recover proper damages or just compensation, or both, from such cable system operator pursuant to the provisions of Title V of this Act, for any injury or other loss sustained by it by reason of such use;

(b) adopt rules or regulations respecting the equal employment opportunities to be afforded by cable system operators;

(c) require the maintenance of such records by cable operators and the submission of such reports to the Commission as may be necessary and relevant to the performance of its duties and responsibilities as provided in this Title or under any rule, regulation, or order adopted pursuant thereto.

Broadcast Signal Transmission and Program Origination

Sec. 205. The Federal Communications Commission may:

(a) make such rules and regulations governing the terms and conditions of the carriage or retransmission of radio and television broadcast signals by cable system operators or channel programmers as may be necessary or reasonably ancillary to the Commission's regulation of radio and television pursuant to Title III of the Communications Act of 1934, as amended, and not otherwise inconsistent with any provision of this Act, taking into consideration any payments received, either directly or indirectly, by the owners of the program material broadcast or the licensees of the broadcast signals, as the result of such carriage, whether by contract, operation of copyright law, or otherwise; *provided, however,* that any such rule or regulation that limits or restricts the number or source of broadcast television signals that may be carried or retransmitted by a cable system, must be based on an evidentiary finding that broadcast service to the public will be significantly diminished in the absence of such rule or regulation.

(b) make such rules and regulations governing pay cable originations of any feature film that has been released and distributed in movie theaters throughout the United States where an admission fee was charged or professional sports event program of any sport regularly televised by a television network, as may be necessary to prevent such programming from being withdrawn from or denied to broadcast television by reason of its carriage on cable television systems, taking into consideration the public interest in fostering competition among television media; *provided* that any such rule or regulation must be based on an evidentiary finding that such feature film or professional sports programming is in fact being withdrawn from or denied to broadcast television by reason of its carriage on cable television systems in the absence of any such restriction, and that substantial segments of the public are being denied the availability of such programming in consequence; such findings to be based on substantial and compelling evidence, and the burden of proof in any evidentiary proceeding initiated by the Commission pursuant to this section to be on the parties seeking to limit or otherwise restrict the availability of such pay cable originations.

Cross Media and Multiple Ownership

Sec. 206(a). The Federal Communications Commission may adopt or continue in force rules or regulations that limit the number of cable systems that may be owned or controlled in common by any person, and that limit or prohibit, as appropriate, the common ownership or control of a cable system, including the facilities specified in subsections 105(a)(1) and 105(a)(2) of this Act, by any television broadcast station or television network, telephone common carrier providing exchange or toll services within the meaning of subsections 3(r) and 3(s) of the Communications Act of 1934, as amended, or by any person having a significant ownership interest in a newspaper or magazine publishing entity, if such television station or network, telephone common carrier, newspaper or magazine publishing entity serves the same area served by such cable system, and if the purpose or effect of such common ownership or control is to substantially lessen competition or restrain interstate commerce, or to restrict the diversity of viewpoints available in the community, *provided that:*

(1) any such telephone common carrier may provide to cable system operators, pursuant to a tariff or other lawful schedule of charges and conditions, transmission facilities used to distribute or disseminate electromagnetic signals from the primary control center of a cable system to cable subscribers;

(2) in adopting or continuing any rule or regulation under this section, or in considering a petition for waiver of any rule or regulation so adopted or continued, that is premised on the public interest in fostering a diversity of viewpoints in the community, the Commission shall consider whether such public interest is met where the operator of such cable system provides only the services specified in Section 303(f), and makes the balance of the channel capacity of such system available for lease to channel programmers having no ownership affiliation with the cable operator.

(b) Nothing contained in this Act shall be deemed in any way to affect the applicability of the antitrust laws of the United States, or the responsibilities of the Department of Justice, the Federal Trade Commission, or the Federal Communications Commission to enforce such laws, as the law may provide.

Administrative Sanctions

Sec. 207(a). When any cable operator (1) has failed to operate substantially as set forth in a certificate of compliance, (2) has violated or failed to observe any provision of Title II of this Act or section 1304, 1343, or 1464 of Title 18 of the United States Code or section 508 and 509 of Title 47 of the United States Code with respect to any program originated by such operator, or (3) has violated or failed to observe any rule, regulation, or order of the Commission authorized by this Act, the Commission may at its discretion:

(1) issue a cease and desist order pursuant to the procedures set forth in section 312(c), (d) and (e) of Title 47 of the United States Code; or

(2) impose in addition to any other penalties provided by law, a forfeiture of not more than $1,000 payable to the United States for each and every day during which such offense occurs or continues, subject to the procedures, conditions, and provisions of sections 503(b)(2) and (3), 504, and 505 of Title 47 of the United States Code.

(b) The Commission may revoke a certificate of compliance, pursuant to the procedures set forth in section 312(c), (d) and (e) of Title 47 of the United States Code—

(1) for false statements knowingly made either in the application for a certificate of compliance or in any related statement of fact submitted with such application;

(2) because of conditions coming to the attention of the Commission which would warrant it in refusing to grant a certificate on an original application;

(3) for willful or repeated violation of, or willful or repeated failure to observe any provision of Title II of this Act or any rule or regulation of the Commission authorized by this Act or by a treaty ratified by the United States; or

(4) for violation of or failure to observe any final cease and desist order issued by the Commission under this section.

Title III. Cable Licensing Program: Authority, Standards, and Requirements

Authority

Sec. 301. A cable licensing authority shall have exclusive authority under state law to execute and enforce the provisions of this Title and to adopt all other rules, regulations, and procedures respecting those activities characteristic of cable system construction and operation as are consistent with the provisions of this Act, and are not exclusively reserved to the Commission by Title II, or forbidden to an executive agency of the United States, a State, or any agency thereof by Title IV of this Act.

Cable License

Sec. 302. No person shall construct or operate a cable system unless such person is issued a license by a cable licensing authority pursuant to the standards and requirements of this Title.

Licensing Standards and Requirements

Sec. 303. Any licensing authority engaged in licensing cable systems shall:

(a) adopt procedures for the issuance, revocation, transfer, or denial of cable system licenses, and for all proceedings incidental thereto, including, but not limited to, procedures providing for adequate public notice of any such proceeding, and providing for public hearing, including the opportunity to submit written comments, prior to disposition of any such proceeding;

(b) adopt procedures providing for the imposition of sanctions upon a finding that the terms and conditions of the cable license have been violated;

(c) grant or renew licenses that are nonexclusive and issued for limited periods of time of no less than five years and no more than twenty years;

(d)   assure that a licensee is qualified to construct and operate a cable system; *provided* that a licensing authority shall not grant a license to any person, including entities under common control, who either directly or indirectly owns or controls access to interconnection facilities serving cable systems and also supplies programming to channel programmers; unless such person certifies that either interconnection services or programming supply services will not be provided to the cable system for which such person seeks a license;

(e)   assure that cable systems constructed or substantially modified after the effective date of this Act are constructed with channel capacity of at least one equivalent channel for every three channels used or intended to be used by the cable operator for retransmission of the signals of television broadcast stations or program originations by such operator;

(f)   require each cable operator with cable system channel capacity in excess of that necessary to transmit (1) the radio and television broadcast signals authorized to be carried by the Commission, (2) a public access television channel as specified in section 401(b), and (3) two additional television channels reserved for program originations by the cable operator, to make available such excess capacity for lease to channel programmers, including independent channel programmers having no ownership affiliation with the cable operator;

(g)   provide in the cable license that excess channel capacity, as defined in subsection 303(f), used for program originations by a cable operator, or any channel programmer owned in whole or in part, or otherwise controlled or operated by such operator, may be diverted from such use and be made available for lease to independent channel programmers having no common ownership or operational affiliation with such operator, upon reasonable demand therefor and pursuant to a formula or other terms and conditions set out in the cable license and reasonably related to the use of, and unfulfilled demand for, such capacity; *provided, however,* that the cable operator shall have available at all times, sufficient channel capacity to transmit the signals specified in subsection 303(f), and may not be compelled to rebuild or reconstruct the cable system or to otherwise increase the total channel capacity of such system;

(h)   assure that each cable operator publishes a schedule of rates, and all changes thereto, setting out the charges, terms, and conditions for the use of channels or time on those channels for program originations, and the access to and use of all instrumentalities, facilities, apparatus, and services incidental to such use of channels or time, which do not unreasonably discriminate among comparable uses or classes of channel programmers, *provided, however,* that nothing herein shall be construed to prevent a cable operator from establishing, as separate classes of channel programmers, persons engaged in educational, eleemosynary, nonprofit, governmental, or similar noncommercial activities, and offering lower rates to such classes of channel programmers;

(i) require the operator of a cable system, who also functions as a channel programmer on such cable system, to establish a separate corporation or entity to perform such function, which corporation or entity shall be accorded no more favorable terms and conditions respecting program originations than are accorded a channel programmer having no ownership affiliation with the cable operator;

(j) assure that the cable system operator does not prohibit the cable subscriber from attaching or connecting to the cable system receiving or terminal equipment of any type, except upon a showing by the operator that the Commission has determined such equipment to be technically incompatible with the operations of a cable system.

Title IV. Limitations on Government Authority

General Limitations

Sec. 401. No executive agency of the United States, including the Commission, and no State or political subdivision or agency thereof, including a cable licensing authority, shall:

(a) require or prohibit program originations by a cable operator or channel programmer, or impose upon such operator or programmer any restrictions or obligations affecting the content of such program originations, including rights of response by any person, opportunities for appearances by candidates for public office, or requirements for balance and objectivity; *provided* that nothing herein shall be deemed to affect the criminal or civil liability of channel programmers pursuant to the laws of libel, slander, obscenity, incitement, invasions of privacy, false or misleading advertising, or other similar laws, except that the cable operator shall not incur such liability for any program originated by a channel programmer having no ownership affiliation with the cable operator;

(b) require the reservation or dedication of cable channels, or time on such channels, to particular persons or uses, except that a cable licensing authority may require the reservation of one standard television channel and require a cable system operator to make time available free of charge on that channel upon the request of any person for any noncommercial or nonprofit purpose, pursuant to such terms and conditions, consistent with subsection (a) of this section, as the cable licensing authority shall by regulation adopt;

(c) establish or adopt specifications respecting the technical characteristics or channel capacity of cable systems, or the technical characteristics of electromagnetic signals disseminated over such systems, except as otherwise provided by section 203(a) or 303(e), or as may be incidental to any rule or regulation adopted by the Commission pursuant to section 205(a);

(d) establish, fix, or otherwise restrict the rates charged channel programmers by cable operators for the use of channels or time on such channels for a period of ten years after the effective date of this Act, or the rates charged advertisers or cable subscribers by any channel programmer for the sale of time or for any program origination; *provided, however,* that the licensing authority may establish reasonable fees, rates, or other charges to be imposed upon cable subscribers by the cable operator for providing services other than program originations to subscribers or for cable system installation, connection, or maintenance at the premises of the subscriber.

Fees

Sec. 402(a). The Commission and the cable licensing authorities of any state, or political subdivision thereof, may collect a reasonable fee from a cable operator upon issuance of a certificate of compliance by the Commission or issuance of a license by a cable licensing authority, *provided* that such fee be based upon reasonable administrative costs to the Commission or such cable licensing authority less that portion of such administrative costs which is sustained for the benefit of the public.

(b) Nothing in this section shall be construed so as to limit the authority or right of any state, political subdivision or agency thereof, or cable licensing authority to contract with any cable operator for the purchase, lease, or other use of public property, rights-of-way, or other public services as may be provided by a state, political subdivision or agency thereof, or cable licensing authority on a contractual basis to other commercial enterprises.

Discriminatory Taxation

Sec. 403(a). As used in this section,

(1) The term "assessment jurisdiction" means a geographical area, such as a State or a county, city, or township within a State,

which is a unit for purposes of determining assessed value of property for ad valorem taxation.

(2) The term "cable system property" means property that is owned or used by any cable system, cable operator, or cable subscriber and is reasonably necessary for the provision of cable services.

(3) The term "commercial and industrial property" means property devoted to a commercial or industrial use, but does not include cable system property.

(4) The term "all other property" means all property, real or personal, other than cable system property.

(b) The following actions by any State, or political subdivision or agency thereof, whether taken pursuant to a constitutional provision, statute, administrative order or practice, or otherwise, constitute an unreasonable and unjust discrimination against and an undue burden upon interstate commerce and are prohibited:

(1) the assessment, for purpose of a property tax levied by any taxing district, of cable system property at a value which, as a ratio of the true market value of the property, is higher than the ratio of assessed value to true market value of all other industrial and commercial property which is in the assessment jurisdiction in which is included the taxing district, and which is subject to a property tax levy;

(2) the collection of any ad valorem property tax on cable system property at a tax rate higher than the tax rate generally applicable to all other commercial and industrial property in the taxing district;

(3) the collection of any tax on the portion of an assessment which is prohibited; and

(4) the imposition of any other tax which results in discriminatory treatment of a cable system, cable operator, or cable subscriber.

(c) If the ratio of assessed value to true market value of all other commercial and industrial property in the assessment jurisdiction cannot be established, then the following actions are also prohibited:

(1) the assessment of cable system property at a value which, as a ratio of the true market value of the property, is higher than the ratio of assessed value to true market value of all other property which is in the assessment jurisdiction in which is included the taxing district, and which is subject to a property tax levy; or

(2) the collection of an ad valorem property tax on cable system property at a tax rate higher than the tax rate generally applicable to all other property in the taxing district.

Title V. Miscellaneous

Right of Action

Sec. 501. Any person adversely affected or aggrieved by any act, practice, or omission of a cable licensing authority or cable operator may bring an action in a court of competent jurisdiction to challenge such act, practice, or omission, on the ground that it does not comply with the provisions of this Act or the provisions of a statute, ordinance, or law of a State, or political subdivision thereof, intended to implement or apply the provisions of this Act.

Federal Court Jurisdiction

Sec. 502. The district courts of the United States shall have jurisdiction of any action commenced pursuant to section 501 and jurisdiction to issue such writs of injunction as may be necessary to restrain the Commission, or any state, or political subdivision or agency thereof, including any cable licensing authority, from violating any provisions of this Act, without regard to the citizenship of the parties or the amount in controversy, and notwithstanding the provision of section 1341 of Title 28 of the United States Code, *provided* that judicial review of actions of the Federal Communications Commission pursuant to this Act shall be in accord with section 402 of the Communications Act of 1934, as amended.

Privacy of Communications

Sec. 503. In order to protect the privacy and security of cable communications, no person shall intercept or receive program originations or other communications provided by means of a cable system unless authorized by the cable operator, the program originator, or other sender of the communication; and no cable operator, or channel programmer, shall disclose personally identifiable information with respect to a cable subscriber or the programming or other communications service provided to or received by a subscriber by means of the cable system except with the consent of the subscriber, or except pursuant to a court order authorizing such disclosure. If a court shall order disclosure, the cable subscriber shall be notified of such order by the cable operator or other person to whom such order may be directed, within a reasonable time before the disclosure is to be made.

Report to the Congress

Sec. 504. The Commission shall submit annually to the Congress a full and comprehensive report on the status of cable communications in the United States, including information pertinent to the achievement of the national policy goals of separating control of cable systems from control of the content of cable channels, together with any recommendations which the Commission may consider appropriate; *provided* that the report required by this section may be made a part of the report required to be submitted by section 4(k) of the Communications Act of 1934, as amended.

Effective Date

Sec. 505. This Act shall be effective eighteen months following its enactment.

APPENDIX C

Draft Cable Television Legislation
Proposed by the Department of Justice, June 1975

Summary and Explanation of the Comprehensive Cable
Communications Act of 1976

The chief problem all draft legislation in the cable area to date has addressed is the need to bound administrative action. It does little good to propose for enactment legislation which will require "reasoned elaboration" by the Federal Communications Commission because of lack of clarity, detail, or direction.

The attached draft bill is by far the most comprehensive draft legislation yet prepared. Included in it are provisions dealing with virtually every point raised in the recent House Communications Subcommittee Report.

The draft bill is comprehensive in the sense that it is not intended as an interstitial amendment to the 1934 Communications Act. The key to understanding the considerable detail of the bill is section 201, which provides that this Act is exclusive of any other authority enjoyed by the Commission under any other law. (The exclusive nature of the bill requires inclusion of the administrative features; the general administrative and organization provisions of the 1934 Communications Act would not allow the Commission to implement the provisions of this new Act.)

The comprehensive nature of this draft bill closely bounds the FCC's potential discretion and flexibility, as it does with regard to State or local officials. Each requirement for public or private action and nearly all rights and obligations conferred are spelled out in detail.

This draft bill is reproduced here in the form and editorial style in which it was prepared by the Department of Justice.

It should be borne in mind that because of the nature of this comprehensive draft bill, regulation of cable television at the Federal level could be shifted to any agency, should that be necessary or desirable in the future, by simple amendment.

This measure was drafted by the Antitrust Division of the Department of Justice and was reviewed by the Office of Legal Counsel and found sufficient in May 1975.

A BILL

To effect a comprehensive National policy respecting cable communications, and for other purposes.

Be it enacted by the Senate and House of Representatives of the United States of America in Congress assembled,

Short Title

Sec. 101. This Act may be cited as the "Comprehensive Cable Communications Act of 1975."

Findings

Sec. 102. The Congress hereby finds—

(a) That cable systems are engaged in interstate commerce through origination, transmission, distribution, and dissemination of television and other electromagnetic signals, and have the technological capability to provide a multiplicity of varied communications services to the people;

(b) That the expansion, development, and regulation of cable communications, while a matter of importance to non-Federal governments is also a matter of appropriate concern to the Federal government;

(c) That the application of policies intended for broadcast communications is inappropriate for cable communications in that cable technology eliminates the channel scarcity found in television broadcasting, and thus facilitates the provision of programming and other communications services not otherwise available over broadcast facilities;

(d) That it is necessary and appropriate for the Federal Government to establish and support a national cable policy to provide a fair opportunity for cable to develop as a medium in its own right, open to all and free from both excessive concentrations of private power and government restrictions that would deny the public the full potential of cable services; and

(e) That a national and uniform policy is needed for cable to prevent the emergence of conflicting or duplicatory regulation by various governmental authorities.

Declaration of Purposes and Policy

Sec. 103. The Congress accordingly declares that the purposes of this Act are to—

(a) Initiate an evolutionary plan, pursuant to its powers to regulate interstate commerce, that will result in the adoption of a comprehensive National policy to allow the growth and development of cable communications, which policy will be responsive to and serve the needs and interests of the public;

(b) Create a regulatory framework, as the first step in such an evolutionary plan, which would apportion the authority to regulate cable systems between the Federal Government and a non-Federal level of government, and which would provide uniform standards and guidelines for the exercise of such regulatory authority by the several states or their political subdivisions;

(c) Assure that cable develops as a communications medium open to all, free of both excessive concentrations of private power and undue government regulation and control that would inhibit the communication of information and ideas, or otherwise deny the public the full benefit of the services to be provided or offered over cable systems;

(d) Assure that cable is provided a fair opportunity to develop as a medium of communications in its own right, allowed to compete freely in the marketplace with other communications media, and is not in relation to such other media limited by regulation to an auxiliary or supplementary role in the provision of communications services to the public;

(e) Establish as the goal of the evolutionary plan initiated by this Act the eventual separation of control of cable systems from control of the content of cable channels, so that such content may be insulated from any local monopoly power of a cable system, as well as from the government regulatory power that otherwise would be necessary to prevent abuses of such monopoly power;

(f) Assure that cable is regulated at the Federal and the non-Federal level of Government in a manner designed to achieve the National policy goal of eventually separating control of cable systems from control of the content of cable channels.

Application of Policy

Sec. 104. The provisions of this Act shall apply as follows:

(a) Any cable system licensed by a cable licensing authority on or after the effective date of this Act shall be subject to the provisions of this Act, and to such rules, regulations, or orders as may be lawfully adopted pursuant to this Act.

(b) Any cable system licensed prior to the effective date of this Act shall be subject to its provisions unless otherwise provided in this Act, and shall be subject to such rules, regulations, or orders as may be lawfully adopted pursuant to this Act, when the cable system's license period ends, or two years from the effective date of this Act, whichever shall occur first.

(c) A State or any duly empowered political subdivision or agency, board, commission, or authority thereof, may adopt or continue in force any law, rule, regulation, order, or standard affecting cable systems; *provided* that such law, rule, regulation, order, or standard is (1) consistent with the exclusive authorities specified in Title II or Title III of this Act; (2) not prohibited by Title IV of this Act; and (3) does not otherwise create an undue burden on the interstate commerce in cable communications.

Definitions

Sec. 105. As used in this Act, and unless the context otherwise requires, the term—

(a) "Cable system" means a facility or combination of facilities under the ownership or control of a single person and authorized to serve a particular geographic area, which consists of a primary control center used to receive and retransmit, store, process, and forward, or to originate radio, television, or other intelligible electromagnetic signals, and transmission facilities with multi-channel capacity, used to distribute such signals over one or more coaxial cables, or other closed or shielded transmission media, from the primary control center to a point of reception at the premises of a cable subscriber. The term "cable system" does not include a facility or combination of facilities that—

(1) serves fewer than 500 subscribers; or

(2) serves only to retransmit the signals of radio and television broadcast stations located within the market in which such facility is located.

A communications common carrier or a miscellaneous common carrier shall not, for the purposes of this Act, be deemed to be a cable system solely by reason of receiving or transporting radio, television, or other intelligible electromagnetic signals in the ordinary course of its business as such a communications common carrier or miscellaneous common carrier.

(b) "Cable channel" or "channel" means that portion of the electromagnetic frequency spectrum used in a cable system for the propagation of a radio, television, or other intelligible electromagnetic signal.

(c) "Multi-channel capacity" means the capacity of a cable system to transmit simultaneously five or more television signals.

(d) "Closed transmission media" means media having the capacity to transmit simultaneously intelligible electromagnetic signals over a common transmission path, and includes media such as coaxial cable, wire, optical fibers, waveguides, and the like.

(e) "Cable operator" or "cable system operator" means any person, or agent or employee thereof, that operates a cable system, or that directly or indirectly owns a significant interest in a cable system; or a person who otherwise controls or is responsible for, through any arrangement, the management or operation of a cable system.

(f) "Cable licensing authority" means any agency, commission, board, or authority of any State, county, municipality, or political subdivision thereof, that is empowered by law to authorize by license, franchise, permit, or other instrument of authority, the construction and operation of a cable system within the jurisdiction of such agency.

(g) "Person" means an individual, partnership, association, joint stock company, trust, or corporation, and for the regulatory purposes of this Act shall include the United States.

(h) "Interconnection facilities" means microwave equipment; boosters, translators, repeaters, communications space satellites, or other apparatus or equipment, used for the relay or transmission and distribution of television, radio, or other intelligible electromagnetic signals to a cable system.

(i) "Cable subscriber" means any person who, for payment of any consideration, receives radio, television, or other intelligible electromagnetic signals distributed or disseminated by a cable operator or a channel programmer over a cable system.

(j) "Channel programmer" means any person who leases, rents, or is otherwise authorized to use the facilities of a cable system for the origination of programming or other communications services over a cable channel, except the use of a channel by a cable subscriber to transmit an electromagnetic signal. Such term shall include a cable system operator to the extent that such operator, or person under common ownership or control with such operator, is engaged in program origination.

(k) "Origination" or "program origination" means the use of a cable channel by a channel programmer to distribute or disseminate any program or other communications service, except retransmission of the signals of a radio or television station by a cable operator or a person under common ownership or control with such operator.

(l) "Cable license" means the license, franchise, permit, or other authorities issued to a cable system by a licensing authority.

(m) "Commission" means the Federal Communications Commission.

(n) "Secretary" means the Secretary of Commerce.

(o) "Market" means a Standard Metropolitan Statistical Area, as defined by the Office of Management and Budget, subject to such modifications with respect to individual areas as the Commission on the motion of any interested party or on its own motion may determine to be appropriate in a particular proceeding for the purposes of this Act.

(p) "Proceeding" includes proceedings as defined in section 551(12) of Title 5, United States Code.

(q) "State" means the District of Columbia, the Commonwealth of Puerto Rico, the Virgin Islands, Guam, Wake Island, Midway Island, the Canal Zone, American Samoa, the Trust Territory of the Pacific Islands, and the Commonwealth of the Northern Mariannas.

Title II. Federal Authorities, Responsibilities, and Functions with Respect to Cable Communications

Federal Communications Commission

Sec. 201. The Federal Communications Commission, created by section 1 of the Communications Act of 1934 (47 U.S.C. §151) shall have jurisdiction and exercise authority with respect to cable communications solely as hereinafter specified. The authorities, responsibilities, and functions of the Commission contained in this Act shall supersede and be in lieu of any other authorities, responsibilities, and functions of the Commission with respect to the regulation of cable communications.

Sec. 202. *Organization and Staff.*

(a) There is hereby created as an agency within the Federal Communications Commission an Office of Cable Communications. (Hereinafter referred to as "the Office.") The Office shall be headed by an Administrator, who shall be appointed by the Chairman, subject to the approval of a majority of the Commission, not later than six months following enactment of this Act. No individual so appointed may receive pay in excess of the annual rate of basic pay in effect for grade GS-18 of the General Schedule.

(b) The Chairman, subject to the approval of a majority of the Commission may employ such other officers and employees (including attorneys) as are necessary in the execution of the functions of the Office, and without regard to the provisions of 47 U.S.C. §155(b), relating to integrated bureaus, but not to exceed fifty, including experts and consultants hired pursuant to 5 U.S.C. §3109, for the fiscal years ending June 30 of 1976, 1977, 1978, and 1979.

(c) Any personnel, property, records, obligations, and commitments which are now used by any bureau of the Commission to regulate cable communications may be transferred by the Chairman to the Office of Cable Communications. The transfer of personnel pursuant to this paragraph shall be accomplished within three years after enactment of this Act and shall be without reduction in classification or compensation for one year after such transfer, except that the Chairman shall have full authority to reassign such personnel during the one year period to other bureaus or offices of the Commission.

(d) Section 5108(c) of Title 5, United States Code, is amended by adding the following new paragraph:

"(15) The Chairman of the Federal Communications Commission, subject to the standards and procedures prescribed by this chapter, may place an additional four positions in the Federal Communications Commission in GS-16, GS-17, and GS-18 for the purposes of carrying out the Cable Communications Act of 1975."

Sec. 203. *Certificates of Compliance.*

(a) No person shall operate a cable system licensed after the effective date of this Act unless such person is issued a certificate of compliance by the Commission.

(b) The Commission shall grant certificates of compliance to any applicant or petitioner upon its showing that it is in substantial compliance with the rules and regulations of the Commission adopted pursuant to this Act, and that it holds a license to construct and operate a cable system issued by a licensing authority pursuant to Title III and Title IV of this Act. The Commission may not in any way condition the granting of any certificate of compliance but may only grant or deny the application for such certificate as filed. No certificate once granted may thereafter be revoked; *provided, however,* that the Commission may require a new certificate of compliance upon (1) the expiration of a franchise, or (2) a substantial change in ownership. In these two instances, a cable system operator may continue to operate its system pending completion of Commission review of its application for a new certificate.

(c) No person holding a certificate of compliance granted prior to the effective date of this Act shall be required to apply for a new certificate simply by reason of the enactment of this Act. No person having applied for a certificate of compliance prior to the effective date of this Act shall be required to submit any further or duplicative information to the Commission, but shall be accorded a speedy determination whether a certificate shall be issued. In no event shall the denial of an application for a certificate of compliance, standing alone, be ground for any subsequent denial or denials.

(d) Applications for a certificate of compliance filed after the effective date of this Act shall be filed with the Administrator. No more than 5 copies of any document shall be required to be filed, and no information shall be required to be filed if previously submitted to the Commission. The Commission by regulation may prescribe reasonable fees to be paid by an applicant for or in connection with the granting of a certificate of compliance, but in no event shall such fees for a certificate of compliance in the aggregate exceed $1,000.

(e) Within 120 days after filing a petition or application for a certificate of compliance, the Commission shall either grant or deny the petition. If the Commission denies a petition (or if it fails to grant or deny such petition within the 120-day period) the petitioner may commence a civil action in a United States district court to compel the Commission to take the action requested. Any such action shall be filed within 60 days after the Commission's denial of the petition, or if the Commission fails to grant or deny the petition, within 60 days after expiration of the 120-day period. *Provided, however,* that the remedies under this section shall be in addition to, and not in lieu of, other remedies provided by law.

Sec. 204. *Broadcast Signal Retransmission; Other Programming.*

(a) The Commission shall not by any single general rule or regulation determine the terms and conditions respecting the retransmission of radio and television broadcast signals by cable system operators or channel programmers, and shall not adopt any single general restriction with regard to the classes or types of programming that may be provided to cable subscribers on a per-channel, per-program, or other basis, except as provided for in this Act. The Commission shall have no authority to determine or to enforce property rights in broadcast programming under Federal or State copyright laws.

(b) The Commission may by rulemaking initiated pursuant to this Act determine the minimum level of broadcast television programming availability necessary in the public interest in markets of specified sizes.

(c) Upon petition of an interested party or recommendation of the Administrator, the Commission may impose requirements or restrictions on the retransmission of television broadcast signals or restrictions with regard to classes or types of programming that may be provided to cable subscribers on a per-channel, per-program, or other basis, by a cable system or systems serving a given market, upon a finding by the Commission that such requirements or restrictions are essential to maintain the minimum level of broadcast service necessary to serve the public interest in that market as determined under subsection 204(b). In any such proceeding, the party or parties urging such requirements or restriction shall bear the burden of proof.

(d) The remedy afforded under subsection 204(c) of this Act shall be in lieu of any other remedy provided for by law, and an order under that subsection may be issued only after an opportunity for a hearing in accordance with section 554 of Title 5, United States Code, except that, if the Commission determines that any person who wishes to participate in such a hearing is a part of a class of participants who share an identity of interest, the Commission may limit such person's participation in such hearing to participation through a single representative designated by such class (or by the Commission if such class fails to designate such a representative).

(e) In any proceeding initiated pursuant to section 204 of this Act, and in all other adjudicatory and rulemaking proceedings under this Act, the Commission shall take into account and consider the public interest to be protected by the antitrust laws and this Nation's general policy of promoting competition and shall adopt the least anticompetitive alternative which will provide the predetermined minimum level of broadcast service in issuing any order, or adopting any rule or regulation.

Sec. 205. *Technical Standards.*

(a) The Commission may by rule promulgate minimum technical standards required to promote the compatibility and interoperability

137

of cable systems, the compatibility of receivers or other terminal equipment connected to such systems by cable subscribers, or to prevent harmful interference to radio communications.

(b) The Commission, in promulgating standards pursuant to this section shall take into consideration the public interest in promoting competition and shall adopt performance as distinguished from design standards wherever feasible.

(c) No person shall manufacture, import, sell, offer for sale or lease, ship, or use devices which fail to comply with regulations promulgated pursuant to this section, subject to subsection 205(b) of this Act. Any person adversely affected by such regulations may petition the Commission for a waiver or variance, and the Commission shall grant or deny any such petition within 120 days after it is filed with the Administrator, stating in writing the reasons for its action. Whenever the Commission grants any such waiver or variance, it shall publish a statement of the reasons for such action in the Federal Register.

Sec. 206. *General Functions and Responsibilities of the Commission.*

(a) The Commission may adopt rules and regulations governing the matters specified below:

(1) *Equal opportunity employment.* The Commission may adopt rules, regulations, and procedures to be followed by cable systems to insure and promote full equal employment opportunity without regard to race, color, religious creed, sex, or ethnic heritage.

(2) *Media diversity and competition.* The Commission may adopt rules or regulations limiting or prohibiting the common ownership of cable systems and television broadcast stations or television networks in the areas served by such stations or networks; rules or regulations limiting or prohibiting the ownership or control of cable systems by persons having a significant ownership interest in a newspaper or magazine publishing activity; or, rules and regulations limiting the number of cable systems any one person may own, nationally, regionally, or in a given market.

(3) *Access to rights-of-way.* The Commission may adopt rules or regulations necessary to assure cable system operators reasonable access at equitable rates to poles, ducts, conduits, and other such rights-of-way owned or controlled by communications common carriers, and electric, gas, and other public utilities, for the purpose of constructing, operating, or maintaining the transmission facilities of a cable system.

(4) *Environmental protection.* The Commission may adopt rules, regulations, or procedures to be followed by cable systems to safe-

guard and to promote the quality of the environment, including the submission of environmental impact or other statements as may be required.

(5) *Federal elections.* The Commission may adopt such rules and regulations regarding the use of cable facilities by candidates for Federal elective office as may be required by the Federal Election Campaign Act, 2 U.S.C. §431, as amended by Public Law 93-443, including rules and regulations regarding the extension of credit to candidates by cable system operators.

(b) The Commission shall adopt rules and regulations governing the matters specified below:

(1) *Subscriber choice.* The Commission shall adopt rules and regulations that require cable system operators or channel programmers to assure that cable subscribers will be afforded adequate information reasonably in advance with respect to the nature of programming to be originated.

(2) *Communications common carrier control.* The Commission shall adopt rules or regulations limiting or prohibiting the ownership or control of cable systems, including such facilities specified in subsection 105(a)(1) and subsection 105(a)(2) of this Act, by telephone common carriers providing exchange or toll services within the meaning of subsections 3(r) and 3(s) of the Communications Act of 1934, as amended, in the area to be served by such cable systems or facilities; *provided* that such communications common carriers may provide to cable system operators pursuant to a tariff or other lawful schedule of charges and conditions, transmission facilities used to distribute or disseminate intelligible electromagnetic signals, from the primary control center of a cable system to cable and subscribers.

(c)(1) Except as provided below in subsection (c)(2), any rules, regulations, or procedures, relating to matters as to which the Commission is granted authority in this Act, which are in effect upon the effective date of this Act, shall remain in effect as if promulgated under this Act.

(2) Any rules, regulations, procedures, or orders, in effect upon the effective day of this Act, relating to the retransmission of broadcast signals or restrictions with respect to classes or type of programming that may be provided to cable subscribers on a per-channel, per-program, or other basis shall become null and void 3 years after the enactment of this Act unless any restriction was imposed pursuant to the standards established in section 204 of this Act.

(3) Any existing rules, regulations, procedures, or orders relating to cable communications as to which the Commission is **not**

authorized to regulate under this Act shall become null and void upon the effective date of this Act.

(d) Any appeal taken from decisions and orders of the Commission pursuant to subsections 206(a) and 206(b) of this Act, and any proceeding to enjoin, set aside, annul, or suspend any rule, regulation, or procedure, taken pursuant to subsection 206(c) of this Act, shall be brought by and in the manner prescribed in section 402(a) of the Communications Act of 1934, as amended (47 U.S.C. §402(a)).

Sec. 207. *Privacy of cable communications.*

(a)(1) No person shall intercept or receive program originations or other cable communications provided by means of a cable system unless specifically authorized to do so by the cable system operator, a channel programmer, other sender of the communication, or as may otherwise be specifically authorized by law.

(2) In order to safeguard the right to privacy and security of cable communications, such cable communications shall be deemed to be "wire communication" within the meaning of subsection 802(1) of Public Law 90-351 (18 U.S.C. §2510(1)).

(3) In the event that there may be any difference between the provisions of this Act and Public Law 90-351, or any regulations promulgated thereunder, it is the intent of the Congress that Public Law 90-351 shall be controlling.

(b) No cable operator or channel programmer shall disclose personally identifiable information with respect to a cable subscriber, or personally identifiable information with respect to the programming or other communications services provided to or received by a particular cable subscriber by way of a cable system, except upon the prior written consent of the subscriber, or pursuant to a lawful court order authorizing such disclosure.

(c) If a court shall lawfully authorize or order disclosure pursuant to subsection 207(b), the cable subscriber shall be notified of such order by the cable operator, or other person to whom such order may be directed, within a reasonable time before the disclosure is made. For the purposes of this subsection, a reasonable period of time shall not be less than 3 calendar days.

(d) The Commission, pursuant to the procedures specified in section 205 of this Act for the adoption of standards may adopt guidelines which may be followed by cable systems to insure the privacy and security of cable communications.

(e) Any cable subscriber whose privacy is violated in contravention of this section 207, shall be entitled to recover civil damages as authorized and in the fashion set forth in section 802 of Public Law 90-351, as amended (18 U.S.C. §2520). This remedy shall be in addition to any other remedy available to such subscriber.

Sec. 208. *Emergency preparedness functions.*

The Administrator, subject to the overall policy guidance and direction of the Commission and other Federal authorities may develop plans, policies, and procedures to be followed by cable systems in an emergency or calamitous visitation including any national emergency type situation. The Administrator shall be principally responsible for the coordination of any such plans, policies, and procedures to assure that they are in consonance with national telecommunications plans and policies developed pursuant to section 606 of the Communications Act of 1934, as amended (47 U.S.C. §606), and other authority.

Sec. 209. *Reports.*

(a) The Administrator shall be principally responsible for the preparation of a report to be submitted by the Commission annually to the Congress which report shall set forth clearly a full and comprehensive report on the status of cable communications in the United States. This report may be included as an appendix or other discrete part of the report required by section 4(k) of the Communications Act of 1934, as amended (47 U.S.C. §154(k)). This report shall include information regarding the progress toward achievement of the purposes of this Act, including the achievement of the national policy goal of eventually separating control of cable systems from control of the content of cable channels.

(b) In order to meet its requirements to report fully and comprehensively, the Commission is authorized, pursuant to this subsection 210(c), to elicit information and data from cable system operators. Each request for information to be reported shall be approved by the Comptroller General in accordance with applicable provisions of Title IV of Public Law 93-153 (44 U.S.C. §3512).

Sec. 210. *Violation of rules, regulations, etc.*

(a) Any person who willfully and knowingly violates any rule, regulation, restriction, condition, or order made or imposed by the Commission under authority of this Act, shall in addition to any other penalties provided by law be punished upon conviction thereof, by a fine of not more than $500 for each and every day during which such offense occurs.

(b) Any cable system operator who—

(A) willfully or repeatedly fails to observe any of the provisions of this Act or of any rule, regulation, or order of the Commission prescribed under the authority of this Act; or

(B) fails to observe any final cease and desist order lawfully issued by the Commission, shall forfeit to the United States a sum not to exceed $1,000. Each day during which such violation occurs shall constitute a separate offense.

(c) No forfeiture liability shall be imposed unless a written notice of apparent liability shall have been issued by the Commission and sent by registered or certified mail to the last known address of the cable system operator. A cable system operator so notified shall be granted an opportunity to show in writing within 30 days after his receipt of the notice why he should not be held liable. A notice of apparent liability issued under this Act shall set forth with reasonable specificity the violation alleged and the date or dates that it occurred.

(d) No forfeiture liability shall attach for any violation occurring more than 1 year prior to the date of the issuance of the notice of apparent liability. In no event shall the forfeiture imposed for the acts or omissions set forth in any notice of apparent liability exceed $100,000.

Sec. 211. *Recovery of forfeitures; petition of Attorney General for equitable relief.*

(a) Forfeitures imposed by the Commission shall be payable to the Treasurer of the United States. Forfeitures shall be recoverable in a civil suit in the name of the United States brought in a district where the cable system operator does business. Any such suit shall be a trial *de novo*. It shall be the duty of the various United States Attorneys, under the direction of the Attorney General of the United States to prosecute for the recovery of forfeitures under this Act. The costs and expenses of such prosecutions shall be paid from the appropriation for the expenses of the courts of the United States.

(b) The district courts of the United States, upon application of the Attorney General alleging a failure to comply with or a substantial violation of any of the provisions of this Act by any person including the Commission, shall have jurisdiction to issue a writ or writs in the nature of mandamus commanding compliance with the provisions of this Act.

Sec. 212. *Appropriations.*

There are authorized to be appropriated to carry out the provisions of this Title II for each fiscal year such sums as the Congress shall deem necessary.

Title III. State and Local Government Authorities, Responsibilities, and Functions with Respect to Cable Communications

Sec. 301. *Cable Licensing Authority.* A cable licensing authority shall have exclusive jurisdiction, consistent with State law, to execute and to enforce the provisions of this Title III.

Sec. 302. *Cable License.*

(a) No person shall construct or operate a cable system after the effective date of this Act unless such person is issued a license by a cable licensing authority pursuant to the standards and requirements of this title.

(b) Any cable system operating on the effective date of this Act pursuant to a license, franchise, permit, or other instrument of authority granted prior to the effective date of this Act shall be deemed to have been licensed under this Act. No such person shall be required to apply for a new cable license simply by reason of the enactment of this Act. Any such application shall only be required as specified in the license, franchise, permit, or other instrument previously granted. At such time as a previously granted franchise may expire or a transfer of ownership, or other event, as specified in the previously granted license, franchise, permit, or other instrument, shall occur, the cable licensing authority shall proceed to consider the issuance of a cable license pursuant to section 303 of this Act.

(c) If, in any State, county, municipality, or political subdivision there shall be no agency, board, or authority empowered by law to issue a license, franchise, permit, or other authority for the construction and operation of a cable system, no cable license shall be required as specified in subsection 302(a). The Commission, in such circumstances shall consider that a cable license has been issued for the purposes of section 203 of this Act.

Sec. 303. Cable licensing authorities shall—

(a) Adopt written procedures for the issuance, denial, or revocation of cable system licenses, and rules and regulations governing cable licensing proceedings including—

(1) Procedures providing for adequate public notice of any such proceeding. For the purposes of this paragraph (1) "adequate public notice" shall mean the publishing in a daily newspaper of general circulation in the community involved of a notice of such proceedings not less than 10 nor more than 30 days prior to such proceeding, or such other measures as the Attorney General of the State may deem satisfactory;

(2) Procedures providing for a public hearing, at a reasonable time and place for any such proceeding. For the purposes of this paragraph (2) a "reasonable time and place" shall include that time and place ordinarily used in the community by the county, township, or parish board to conduct public business, or such other time and place as the Attorney General of the State may deem satisfactory;

(3) Procedures providing for the right of citizens of the community to speak and to submit written comments. For the purposes of this paragraph (3) a cable licensing authority may use procedures similar to those used with respect to the discussion of proposed municipal legislation, or such other procedures as the Attorney General of the State may deem satisfactory.

(b) Adopt written procedures with respect to the steps to be taken, if any, including the imposition of sanctions, upon a finding that material and significant terms and conditions set forth in the cable license have been violated.

(c) Grant or renew cable licenses that are (1) nonexclusive; and (2) for a period not less than 5 years nor more than 20 years.

(d) Assure that a proposed cable licensee is qualified to construct and operate a cable system. A cable licensing authority shall not grant a cable license to any person, including persons under common control, who either directly or indirectly owns or controls access to interconnection facilities serving cable systems and who also supplies programming to channel programmers. However, a cable licensing authority may grant a cable license to such person if he shall certify that either interconnection services or programming services, but not both services, will be provided to the cable system for which such person seeks a license.

(e) Assure that cable systems constructed or substantially modified after the effective date of this Act are constructed with adequate channel capacity. For the purposes of this subsection 303(e), the term "adequate channel capacity" shall mean the capacity to transmit simultaneously not less than 12 television signals.

Sec. 304. *Separations policy goal; three factor formula.*

(a) The cable licensing authority shall set forth in the cable license the requirement that the cable system operator make available for lease to independent channel programmers prime time viewing channel capacity, as computed annually in accordance with the three factor formula set forth in this section. The minimum percentage of prime time viewing capacity to be made available shall be the equivalent of the product of the subscriber income factor and the market acceptance factor, expressed as a percentage. For the purposes of this section:

(1) *Subscriber Income Factor.* The subscriber income factor is a fraction, the numerator of which is the per capita income of the area in which the cable system is located and the denominator of which is the per capita income of the State in which the cable system is located. The area per capita income shall be the per capita income of the county (or any smaller geographic area over which the cable licensing authority has jurisdiction) as most recently available from the Bureau of the Census or the Department of Commerce, and used for the purposes of section 109 of Public Law 92-512 (31 U.S.C. §1228). The State per capita income shall be the per capita income of the State in which the cable system is located, as most recently available from the Bureau of the Census or the Department of Commerce, and used for the purposes of section 109 of Public Law 95-512 (31 U.S.C. §1228).

In the event that the cable system shall serve (or propose to serve) more than one county or more than one State, the lower fraction derived from using the relative per capita income figures shall be used.

(2) *Market Acceptance Factor.* The market acceptance factor shall be a fraction, the numerator of which is the number of households subscribing to a cable system and the denominator of which is the number of households within the area served by the cable system. For the purposes of establishing this denominator, the most recently available figures from the Bureau of the Census or the Department of Commerce shall be used.

(3) *Prime Time Viewing Capacity.* Prime time viewing capacity shall be the capacity of a cable system available for lease during the hours 7:00 P.M. through 11:00 P.M., Eastern, Pacific, 6:00 P.M. through 10:00 P.M. Central or Mountain Standard Time; subject to such rules and regulations as may be adopted by the Secretary of Transportation. For the purposes of this paragraph, all channels on a cable system shall be deemed to be available for lease, except those channels used to (1) retransmit the signals of the television stations located within the market in which the cable system is located; or (2) retransmit television signals pursuant to Commission directive under section 204 of this Act.

(b)(1) The cable licensing authority shall, in the cable license, set a date certain, and on that date every year of the franchise, beginning one year after the franchise is issued (or one year after the effective date of this Act, whichever occurs later), the cable licensing authority and the cable system operator shall meet for the purposes of computing the percentage of prime time viewing capacity which shall be made available for lease for the following year.

(2) The cable licensing authority shall be responsible for computing and reporting the Subscriber Income Factor.

(3) The cable system operator shall be responsible for computing the Market Acceptance Factor.

(4) The percentage figure so derived shall be rounded to the next highest number. The cable system operator must make available for lease at least the prime time viewing channel capacity required by this section in each two-week period during the year that follows.

(c) In the event that the cable licensing authority and the cable system operator shall be unable to reach agreement as to the amount of prime time viewing channel capacity required to be made available for lease under this section, such dispute shall be resolved pursuant to the provisions of section 307 of this Act.

Sec. 305. *Cable system services; requirements.*

(a) The cable licensing authority will require in the cable license that the cable system operator publishes and otherwise makes available to the public full information regarding the cable system's current rates, charges, and associated terms and conditions of services. For the purposes of this subsection 305(a), information filed and kept current with the cable licensing authority or other public agency will be considered to have been published.

(b) The cable licensing authority will require in the cable license that the cable system operator makes available to the public full information regarding the availability of channel capacity for lease, and the rates, charges, terms, and conditions for the use of such channel capacity.

(c) The cable licensing authority will require that the cable system operator provide any channel capacity available for lease, and any associated facilities, without discriminating among comparable uses or classes of channel programmers or lessees.

(d) The cable licensing authority may not require that the cable system operator provide free service to any person, or afford any person the opportunity to make use of the cable system or associated equipment for the purpose of originating programming. The cable system operator may establish as a separate class or classes of channel programmers persons engaged in educational, eleemosynary, nonprofit, governmental, or similar noncommercial activities, and may offer lower rates to such classes of channel programmers.

(e) The cable licensing authority will require a cable system operator who desires to function as a channel programmer on its own cable system to establish a separate corporation or entity to perform as such a channel programmer. In such an event, the cable licensing authority will require that the affiliated channel programmer be accorded no more favorable terms and conditions with respect to program origination than those accorded channel programmers not affiliated with the cable operator.

(f) The cable licensing authority will set forth in the cable license the requirement that the cable system operator shall not prohibit the cable subscriber from attaching or connecting to the cable system receiving or terminal equipment of any type, except upon a showing by the cable system operator that the Commission has determined such equipment to be technically incompatible with the operation of a cable system.

Sec. 306. *Rate Regulation.*

A cable licensing authority may not regulate or otherwise endeavor to control the rates charged for or in connection with cable communications services except that ceilings on any charges may be imposed upon cable systems located in markets in which 2 or fewer television broadcast stations are located.

Sec. 307. *Disputes.*

(a) Disputes between a cable licensing authority and a cable system operator arising out of any provision of this Title III shall be settled in accordance with the State's rules of arbitration in effect at the time the dispute arose, and judgment upon the award rendered by the arbitrator(s) may be entered in any Court having jurisdiction thereof. If State law does not provide any means for the arbitration of disputes, then any dispute dealt with in this subsection shall be settled in accordance with the State's laws relating to the adjudication of civil suits.

(b) Disputes between a cable system operator and a cable subscriber or channel programmer arising out of the provisions of this Title III shall be settled in accordance with State laws relating to the adjudication of civil suits.

Title IV. Limitations on Government Authority

Sec. 401. No executive agency of the United States, or the Commismission, and no State or political subdivision or agency thereof, including a cable licensing authority shall—

(a) Require or prohibit program originations by a cable operator or channel programmer.

(b) Restrict or impose obligations with respect to the content of programming that is originated such as any governmentally imposed obligation that channel programming be fair, balanced, or objective; or that persons be afforded rights of response, or that candidates for public office be given opportunities to appear. Nothing herein shall be deemed to affect any civil or criminal liability of a channel programmer for violation of the laws of libel, slander, obscenity, incitement, invasions of privacy, false or misleading advertising, copyright, or other similar laws, except that a cable system operator shall not be held liable by reason of programming originated by a channel programmer not affiliated with the cable system operator.

(c) Impose any special tax or other revenue raising measure upon cable system operators, channel programmers, or cable subscribers solely by reason of the operation or use of a cable system; *provided* that nothing shall be deemed to preclude the imposition of any general sales or excise tax on the sale of cable services; and *provided further* that cable licensing authorities may impose reasonable fees designed to reimburse such authorities for the direct costs, if any, entailed in issuing cable system licenses.

Sec. 402. *Continued Applicability of the Antitrust Laws.*

Nothing contained in this Act shall in any way affect the applicability of the federal antitrust laws.

Title V. Miscellaneous Provisions

Sec. 501. *Separability.* If any provision of this Act, or the application of such provision to any person or circumstance shall be held invalid, the remainder of this Act, or the application of such provision to persons or circumstances other than those as to which it is held invalid, shall not be affected thereby.

Sec. 502. *Effective date.* This Act shall be effective 18 months following the date it shall become law.

APPENDIX D

"Pay Cable Options Paper" submitted to
Paul W. MacAvoy, December 1975

Executive Summary

The FCC's pay-cable rules have the effect of keeping television network costs for prime programming down. They function in essence
as a subsidy to major television advertisers; they also tend to deter
particularly the twenty-five firms that accounted last year for about
one-half of all television revenues from shifting their buying habits
towards greater reliance on local, as distinguished from national,
advertising. In the short term, the FCC rules prevent cable television
subscribers—about 12 percent of the population involved—from buying what they might want to see. By "stabilizing" the demand for
movies, the rules have the effect of making any substantial increase
in program supply, and any resulting increase in potential diversity,
less likely.

The FCC acknowledges that the rules are not designed to prevent
siphoning of movies for free television in any absolute sense. Rather,
as the FCC has stated, the movies rules "only" guarantee that television will have a "timely opportunity" to show these films. That is,
they would in the ordinary course of marketing reach free television
eventually; the commission believes it should act to accelerate that
process.

Sports issues are somewhat different since the leagues act, with
various antitrust exemptions, to curtail the supply of events and their
availability, quite apart from whatever the FCC may do. The need
for FCC meddling, however, is problematical, since the Congress has
a very plain, demonstrated capacity to move very quickly to assure

This paper was submitted as a memorandum to Paul W. MacAvoy from Deputy
Assistant Attorney General Jonathan C. Rose of the Antitrust Division of the
Department of Justice.

the showing of sports on television. Sports siphoning, which the leagues deny would happen, would in any event only occur over time and Congress could certainly act to prevent it.

Probably the best way to remedy the present situation would be to rescind the FCC movies and sports rules outright. A more acceptable approach to the Congress would involve outright repeal, with the caveat that the FCC could act upon a finding that siphoning had actually occurred to the detriment of the public.

Comprehensive Analysis

Background. Since 1970, the Federal Communications Commission's rules have forbidden people who subscribe to cable television from paying additional amounts to watch movies or sports programs that are not available to them on free television locally or by way of "imported" distant television signals. In essence these FCC pay-cable rules bar cable television operators or third parties who lease channels on a cable system from offering subscribers the option of watching movies that are three to ten years old or sports events televised in the last five years on a per-program or per-channel pay basis. These restrictions are imposed as a condition for the cable television system to retransmit any broadcast television signal or signals. There are some exceptions to these rules, but as a practical matter they are not decisionally important.[1]

The FCC pay-cable rules are under review in an appeal in which the Department of Justice is a statutory respondent. Oral argument before the Court of Appeals for the District of Columbia Circuit is tentatively slated for spring 1976.

The basic rationale for the FCC's pay-cable rules is the need to prevent "program siphoning," as distinguished from "audience siphoning," which the FCC's distant-signal rules are designed to prevent. Cable television, it is argued, could develop the market power to divert or "siphon" desirable programs from the three national networks or from local television stations. This is so because advertisers are not willing to contribute as much per viewer attracted by certain programs as those viewers might pay themselves to watch the program on pay cable. Though the number of cable subscribers currently is small—and, indeed, will remain relatively so—their combined pur-

---

[1] For instance, the pay-cable rules do not literally apply to "hotelvision" systems, which show only movies. In response to National Cable Television Association assertions that such systems could siphon *cable* revenues, the FCC imposed constraints on such systems equivalent to the pay-cable rules.

150

chasing power is more than sufficient, it is said, to be able to outbid "free TV." [2]

Not everyone, moreover, has or will have access to cable television services, the FCC maintains. In the flamboyant rhetoric that has accompanied this public policy issue, the "turnstile in the homes of the privileged" will bid away the most desirable sports and movies at the expense of residents of "sparsely settled ... low density areas," [3] the "less affluent," [4] and especially minorities. [5] Unable to provide the revenues needed to sustain "sustaining" (that is, nonprofitable) public service programming, it is argued, the basis for the FCC's traditional policy of "local service" by television "fiduciaries" will be irremediably undermined.

The FCC has never stated that "siphoning" has occurred or is likely to occur. Rather, the FCC has said that the public interest in the quality and the quantity of existing television programming is so great that the "mere potential" of "siphoning" is sufficient to warrant prophylactic action. In its opinion, upon reconsideration of its most recent pay-cable rules, the FCC conceded that its rules regarding movies were aimed not at siphoning in the absolute sense but rather at assuring the "timely exposure" of movies on conventional television. The parties appealing those rules argue that this is a "belated theory of 'delay' " that is deficient as a matter of law.

The FCC has, essentially, declared that the "public interest" standard of the 1934 Communications Act and the commercial prac-

---

[2] In 1973, per-household expenditures for television advertising amounted to an equivalent of sixty-seven dollars per household. Computed from 1974 *Statistical Abstracts*, Tables 50, 1301. Not included is the amount invested by the public in receivers, in watching time, electricity, et cetera. It is argued, however, that "free TV is more than free, because it makes possible mass production and lower costs." Comments of the Association of Maximum Service Telecasters on Pay Cable Rules, in FCC Docket 19554, October 1974, p. 24.

[3] See AMST Comments, pp. 10-18. At this time, 1.2 million American households are not receiving adequate broadcast television service on any channel, either from originating broadcasters, translators, or cable systems. Moreover, 2.4 million households are afforded only one channel; 2.2 million, only two channels. Denver Research Institute, *Broadband Communications in Rural Areas: National Cost Estimates and Case Study* (Denver: University of Denver, 1974).

[4] Comments of the American Broadcasting Company on Pay Cable Rules, in FCC Docket 19554, October 1974, p. 14.

[5] See, for example, ABC Comments, p. 14. In contrast, the National Black Media Coalition has argued that a relaxation of the pay-cable and related rules would promote opportunities for blacks, afford them a chance to view movies they might otherwise be unable to afford, et cetera. See Comments of the National Black Media on Pay Cable Rules, in FCC Docket 19554, October 1974, pp. 2-5. It is noteworthy that the cable penetration in many impoverished areas is said to exceed that of telephone service, and the demand for cable service is less elastic.

tices of the existing system of television broadcasting are one and the same thing. This system of broadcasting does, clearly, provide very substantial benefits to the American public. It is estimated that the "value of free TV to Americans is over $20 billion a year—$25 per TV household" a month in terms of providing entertainment.[6] Roger Noll, for example, has argued that "to pay for exactly the same programming now available without charge would mean a massive reduction in the welfare of most families."[7]

There is some basis in the demographics of cable television operation for the assertion by established broadcasters that cable will focus its service offerings on more lucrative neighborhoods, and that it will not wire less advantaged areas.[8] By the same token, the concentration and resulting profitability of U.S. television broadcasting is such[9] as to discount substantially assertions that cable television, much less pay-cable services, "threatens" the viability of over-the-air broadcasting in any realistic sense. Indeed, the consumer surveys commissioned by the National Association of Broadcasters indicate little public demand for pay-cable services. The most recent Roper Poll

---

[6] Comments of the National Association of Broadcasters, in FCC Docket 19554, October 1974, p. 8, citing R. G. Noll, M. J. Peck, and J. J. McGowan, *Economic Aspects of Television Regulation* (Washington, D.C.: The Brookings Institution, 1973).

[7] Comments of Roger Noll, in FCC Docket 19554, October 1974, p. 1.

[8] Subscriber statistics indicate, for example, that if markets are defined in terms of thirty-five-mile radii, as they are for the purposes of the FCC's cable television rules, there are about 3 million cable subscribers in the top 100 markets. If, however, the broadcasting industry's definition of a market is used, indications are there are about 5.5 million cable subscribers in the top 100 ADIs (areas of dominant influence). The 2.5 million subscriber differential can reasonably be considered suburban in nature; in other words, nearly a quarter of the cable industry's present subscribers live not in rural hamlets, as frequently asserted, but in suburban areas immediately surrounding the largest urban areas. See Ralph Thomlinson, *Urban Structure: The Social and Spatial Character of Cities* (New York: Random House, 1969), p. 102ff.

[9] About one-half of the U.S. television homes are located in the top twenty ADIs; three-fourths are located in the top fifty. About two-thirds of the total television industry profits derive from the top twenty ADIs; 90 percent, according to S. M. Besen, derive from the top fifty. VHF network-affiliated television stations in the top fifty ADIs routinely return thirty cents net income per dollar of sales; in many instances, such returns on revenues are substantially higher. Appraisals of the profitability of television broadcasting in major markets vary widely, as do accounting practices in the industry, though all parties agree it is very substantial. The demonstrated ability of the networks to pay staggering amounts for particular films is described in the department's October 1974 Pay-Cable Comments to the FCC. In fairness, a substantial number of cable television systems routinely return 25 percent and up, after taxes, on equity. The consistent and high profitability of such "old systems" (that is, pre-1968 or 1970) is somewhat obscured by some spectacular lack of profitability of some major cable holding companies.

undertaken in behalf of the Television Information Office found more than 70 percent of those polled in 1974 "not interested" in the option of paying $1 to $2.50 per program, $20 to $40 to $50 per month to see "Broadway shows, the newest movies, championship fights, operas, and other special programs you rarely see now." This may be a function of the existing program offerings mandated by the FCC rules; each of the sample program types, for instance, could be shown on pay-cable right now. Nonetheless, even the "most optimistic projections of the size of the STV (that is, pay-cable or subscription TV) market still leave the vast majority of Americans not subscribing."[10]

Industry Positions. The positions of five industries connected with cable television are summarized as follows:

Cable television industry. The cable television industry representatives' arguments advanced Thursday, October 9, 1975, and amplified in subsequent submissions of the National Cable Television Association are, in essence, that cable's ability to offer certain pay-cable services is probably necessary in order to wire the major markets, at least in the nearer term. This probably is because for the 70 percent of the American population living on 3 percent of the land, five channels or more of television are now available to them over-the-air for free. Construction costs, and hence revenue requirements, for these areas are high. Without the ability to offer "more," cable system owners maintain, their ability to attract and hold these urban subscribers is diminished. Pay-cable services, they argue, are more directly responsive to consumer desires than advertiser-selected and funded programming. The advent of domestic satellite service capable of linking cable systems on a national or regional basis may enhance the marketing potential for pay-cable services.[11]

Cable system operators also argue that the FCC's rules on the kinds of programs that can be offered on a pay basis constitute an unconstitutional prior restraint on free expression. Even assuming that some restraints may be justified, they reason, the present FCC

---

[10] Noll Comments, p. 27.

[11] A feature of the existing FCC pay-cable rules makes the commercial viability of such multimarket link-ups problematical. The rules do allow cable systems to bid for the right to show movies under contract to local stations or to national networks with an affiliate in the market. The rationale for this exception is that cable cannot siphon what free television already has. Unfortunately, the FCC rules are keyed to a market-by-market "clearance"; in addition, the FCC did not operationally take into account the fact that the bulk of movie contracts with broadcasting interests provide for very lengthy "exclusivity" periods.

153

rules lack the firm basis in fact and the reasoned specificity necessary to any government intrusion into an area such as program content that is "fraught with first amendment sensitivities."

The cable system operators did not raise past arguments that, for example, absent sufficient revenues such as pay cable would provide, cable systems cannot be expected to provide the "special public services" made possible by the "promise of cable." They labeled as "scare tactics" assertions that cable television would siphon away sports events, especially name events like the World Series.

Conventional television industry. Representatives of major television broadcasting interests concentrated their interests at the Thursday, October 16, 1975, meeting upon discerning the strategic purposes they saw underlying the issue of cable television "de-regulation."[12] Very little credible data was produced at that time, although subsequent submissions have made the broadcasting industry's case for continued restrictions upon cable and pay cable considerably more persuasive.[13]

The essence of the television industry's argument is that pay-cable television is nothing more than an effort by a "basically parasitical industry"[14] to gain "frosting" for a cake that they have stolen from free television. Cable, they argue has established itself artificially, but wrongfully expropriating the signals of broadcast television, selling those signals to the public, and not sharing the dividends with broadcasters or copyright holders. Pay cable, they argue, is thus

---

[12] For a lengthy discussion of the "ulterior motives" perceived by Columbia Broadcasting System President Arthur Taylor, see Thomas Whiteside's "Annals of Television: Shaking the Tree," *New Yorker*, March 17, 1975. It is noteworthy in this respect that both economic and legal literature reflects recurrently the proposition that to the extent certain firms (or industries) may enjoy market dominance, if not outright monopoly, the firms are as a result relieved of the "burdens of economics." See Richard A. Posner, "Natural Monopoly and Its Regulation," *Stanford Law Review*, vol. 21 (1969), pp. 548, 559, 575-76; Richard G. Lipsey and Peter O. Steiner, *Economics*, 2nd ed. (New York: Harper and Row, 1969), p. 369.

[13] The single assertion of economic "fact" intended to support the television industry's views that cable growth imperils small broadcasters was made by Mr. Carey at the October 16 meeting. Mr. Carey stated that his station had lost "40% of an advertiser's business" because of cable. He did not state which advertiser, what business, or offer any substantiating data during or after the meeting. Mr. Carey's difficulties, and additional information on them, are described at p. 3 of the economic analysis forwarded by ABC to you November 7, 1975.

[14] Statement of C. Wrede Petersmeyer, chairman of Dow Jones' Corinthian Broadcasting, West Lobby, October 16, 1975.

a marginal-cost, high-profit business "cablevision" would engraft on broadcasting's rootstock.[15]

If the already too lax FCC rules are further relaxed, moreover, siphoning is inevitable, broadcasters say. For example, they argue, free television currently pays about $800,000 for a network showing of a movie. They reason that 1.5 million pay-cable homes paying one dollar can outbid free television. If there are 30 million homes wired for cable and pay-cable services, as some cable industry analysts have projected, 5 percent of those homes paying one dollar, or 10 percent paying fifty cents, or 20 percent paying twenty-five cents can generate the funds necessary to outbid free television.[16] Pay cable will focus its buying power, they add, on the most desirable sporting events. The upshot of these developments, they maintain, will be lower quality television and lower revenues to sustain broadcasting's public service obligations. This will be to the especial injury of the economically and geographically disadvantaged segment of the population that will not have cable television.

The television industry does not seek protection from fair competition, they argue. The critical issue is "the quality of free TV service to the public—not injury to broadcasters."[17] By the same token, they maintain that cable television in any form will have, and indeed has already had, a disastrous impact upon small-market broadcasters.

The cable industry, broadcast representatives asserted, should, if it must offer pay-cable services, focus its attentions on providing the specialized kinds of programming not offered by conventional broadcasting. Such programming, it is suggested, could be culturally enriching and could enhance the diversity of programming options available to the public.

Movie producers and distributors. Representatives of the Motion Picture Association of America and major studios stated their position on the FCC's pay-cable rules Monday, October 20, 1975. They argued, basically, that siphoning is a phony issue which has been

---

[15] It may be noteworthy that 34 percent of all cable systems currently are owned by television interests, other than national television network interests (who were directed by the FCC to divest all of their cable holdings in 1970-1971). Only about 100 of these systems and television stations are co-located. Notwithstanding the broadcasters substantial investment in cable television, V. T. Wasilweski, president of the National Association of Broadcasters stated on October 16, 1975, that "every element of this industry is opposed" to any relaxation of the present FCC pay-cable rules.

[16] ABC Comments, p. 8.

[17] AMST Comments, p. 24.

devised by the FCC and established broadcasters to limit competition for their products—movies and some television entertainment programming. MPAA also argued that the FCC's pay-cable regulations were especially objectionable since they represented regulation by the FCC of the way in which movie producers and distributors chose to market their products to maximize their profits.

The movie producers argued that the supply of movies in some ways can be viewed as limitless, though they conceded that not all movies are fungible. With pay cable, they argue, they would sequentially sell their product based upon expectable per capita receipts: three dollars (first-run theaters); two dollars (neighborhood theaters); fifty cents (pay cable); then five cents (network and local television).

The movie producers criticized the national television networks for arguing that the movies rules that the FCC has adopted are desirable, given that the networks have been steadily curbing their appetite for movies in recent years.

Professional sports leagues. Representatives of the professional sports leagues,[18] which we met with Thursday, October 23, and Friday, October 31, 1975, were more guardedly optimistic with respect to the advantages pay cable could offer them. Their position, basically, was that pay cable could offer them a much broader market in which to sell their product than conventional television now affords. Only professional baseball supplied specific figures regarding their revenues from conventional television.[19]

Sports interests stressed the difference between their product and other possible pay-cable fare, arguing that its value is mostly transitory in nature: "The fifth game is worthless since Cincinnati won the Series." Each sports league representative whom we met with explicitly denied any intention of shifting name events such as the World Series to a pay basis. The representative of the National Hockey League pointed out the "double twist" feature of the FCC's "highwater mark rule"—that is, under the FCC's rules, the more of an event free television may televise, the less of the same event pay cable may sell; *and*, the less of an event free television may televise, the less of the same event pay cable may sell.

---

[18] Not including professional football.

[19] Baseball Commissioner Bowie Kuhn stated that baseball revenues from conventional television run about $750,000 per year per club (twenty-six clubs); they are a substantial factor in club income, which averages $6 million a year. For baseball as a whole, television revenues run about 26 percent of gross, 12 percent of gross being purely local. He tended to discount the team loyalty factor as decisive, pointing out that the networks had found baseball games to have substantial national audience appeal.

Each of the sports representatives, however, seemed concerned with the possible "gate siphoning" effects pay cable might have, especially with regard to league members whose poor performance had diminished their gate potential to begin with. The same "box office impact" argument was made more forcefully regarding the prospect of unregulated distant-signal importation.

Theater Owners. National Association of Theatre Owners (NATO) representatives opened the Friday, October 24, 1975, meeting with the somewhat extravagant claim that unregulated pay cable threatened $5 billion in investment and 500,000 jobs.

Subsequent arguments centered on the allegation that the movie industry is currently monopsonistic in nature—very much a producer's market—and that the major producers prefer to focus their marketing efforts on high-price, high-volume theaters in a few major markets. For example, they noted that movies such as *Jaws* were released to only fifty-seven theaters in the country, and that they were not available to other theaters at any price. Examples were given of the extremely high guarantees, payable in advance, required by the majors; that, coupled with rising general-operating costs, made theater recovery of "the nut" (that is, overhead costs per film showing) very difficult, and cut into profit margins severely. Pay cable, they argued, could completely destroy theaters as a profitable business.

The theater owners repeatedly complained that the Antitrust Division, which regulates the major producers under certain consent decrees, had failed to take what they believed was the necessary action to correct the abuses which they perceived. They stated, however, that they would not object to pay-cable deregulation so long as their exclusive right to the "first fruits" from movies was retained.

Nonindustry Representatives. Included here are the positions of public interest groups and academic economists.

Public interest groups. No representative of any public interest group that appeared at the Tuesday, October 28, 1975, meeting expressed a desire to be protected from the alleged harm of siphoning. Each representative favored substantial relaxation, if not abolition, of the FCC pay-cable rules.[20]

---

[20] "There is no justification for a Federal agency adopting rules with the sole purpose of protecing one private interest from another private interest. There is no evidence in the record that the rules do anything except that." Comments of the American Civil Liberties Union on Pay Cable Rules, in FCC Docket 19554, October 1974, p. 1, accord, Comments of the Americans for Democratic Action.

Academic economists. All of the academic economists who appeared at the Friday, November 21, 1975, afternoon discussion were in favor of rescinding the pay-cable rules as they applied to movies and series programs.[21] The rationale advanced by Roger Noll was that the supply of feature films is elastic.

The academic economists were less favorably inclined towards proposals to rescind the FCC pay-cable rules as they pertain to sports events. The rationale that they advanced for possibly retaining some FCC regulations in this area was that sports as such is in some ways monopolistic. The sports leagues have, and exercise, the power to affect the supply of games and the televising of games. This rationale is not dissimilar from that put forward by FCC Chairman Richard E. Wiley, that in requiring the sports leagues to sell, essentially, only to free television, the FCC was doing little more than holding the leagues to the bargains struck in return for various exemptions from the antitrust laws obtained or contemplated over the years.

Congressional Staff. The positions of the staffs of three congressional subcommittees and of the legislative assistant of the Senate minority leader are summarized as follows:

House Subcommittee on Communications. Repeal of the FCC's existing pay-cable rules through legislation was not a major topic discussed at the Wednesday, September 24, 1975, meeting with the subcommittee's staff.

No inference was conveyed that the staff believed there was any public interest purpose underlying these FCC rules. It was stated that the Congress could act very swiftly to correct any siphoning in the sports field, should it occur; privately, members of the staff have expressed a willingness to concede the sports issue altogether in order to make enactment of a complete repeal of the rules as they pertain to movies easier. No sympathy was expressed for the broadcast industry; rather the discussion could be fairly categorized as "pro cable" in almost all respects.

The subcommittee staff stated their intention to prepare comprehensive cable legislation dealing with the pay-cable issue and other issues, and their intention to undertake hearings into cable regulation in the spring of 1976.

---

[21] The series rules, which previously prohibited pay cable from airing programs with an "interrelated plot," et cetera, have been recently placed in abeyance by the FCC, on the ground that there was no evidence in the record that they were necessary.

Senate Subcommittee on Communications. Majority staff of the subcommittee expressed only a willingness to consider convening hearings on cable television legislation in the Monday, September 29, 1975, meeting with them. No threshold biases towards any particular side of the controversy were apparent. Indeed, given that the chairman of the subcommittee is frequently said to be the motivating force behind the FCC's cable rules generally, the neutrality on the issues was somewhat surprising. Minority staff of the subcommittee, however, have apparently stated privately that any proposal to rescind or relax the FCC restrictions on pay-cable carriage of movies is foredoomed to failure.

Senate Subcommittee on Antitrust and Monopoly. The Friday, October 24, 1975, meeting with subcommittee staff was almost purely informational in nature. Some reluctance to tackle sports issues in any fashion was expressed by Mr. Howard E. O'Leary, with reference in particular to the difficulties that nonprofessional college and high school sports create. Subcommittee staff stated that the record of their recently concluded hearings into the competitive effects of the FCC's pay-cable rules and related national television network exclusivity practices supported contentions that the FCC's rules were designed primarily to protect established television interests from new competition.

Subsequent to the meeting, the subcommittee announced that it would prepare legislation to prohibit the FCC from adopting any regulations with respect to pay cablecasting of movies. The subcommittee did not, however, state when such legislation would be ready for introduction.

Mr. Dennis Unkovic. Mr. Dennis Unkovic, previously the legislative assistant of the Senate minority leader, focused most of his remarks during the Thursday, October 30, 1975, meeting on the asserted inadequacies of the pending copyright legislation.

He repeatedly attributed great lobbying and persuasive powers to the NCTA, and he discounted the effectiveness of the NAB or broadcasters generally. At the same time he repeated instances in which broadcasters, through the splicing of film clips for local newscasts, could favor certain politicians. Mr. Unkovic stated that legislation aimed at deregulating cable television would be feasible, given the support of four or five important senators, and "giving the broadcasters something."

Other Organized Groups. The positions of state and local government agencies and of organized labor are summarized as follows:

State/local government agencies. The New York City official responsible for regulating that city's cable industry strongly supported measures that might promote cable television, and strongly criticized the city's broadcasters for "extracting" $40 million in profits annually. The representative of the Minneapolis-St. Paul regional planning group supported measures aimed at relaxing the FCC's cable regulations generally, noting that in their poorer neighborhoods more homes had chosen to subscribe to cable than to telephone service. The city attorney for San Diego noted that while his city was one of the most heavily "wired" cities in the nation, the profitability of area broadcasters, and especially UHF broadcasters, had risen steadily. He stated that in all likelihood the impact of removing the FCC pay-cable rules entirely would be minimal.

Organized labor. No major labor union has apparently stated a position favoring or disfavoring the present FCC pay-cable rules. The past president of the Communications Workers of America, AFL-CIO, expressed personal concern that the public not be required to pay for what they now get for free. The International Ladies Garment Workers, AFL-CIO, signed a brief submitted under AMST auspices challenging the FCC pay-cable rules for being too weak. The Screen Actors Guild has sided with the motion picture industry as favoring repeal of the pay-cable rules as they apply to movies and feature films. Labor unions directly involved in television broadcasting—the cable television industry is essentially unorganized at this time—have not directly stated positions or, if they have, have done so in a relatively invisible fashion.[22] It is unlikely, however, that any of these unions would support significant cable television deregulation.

Who Gains, Who Loses? Who gains and who loses from the FCC pay-cable rules? The three national television networks gain quite a bit. The rules tend to assure them continuance of a noncompetitive, buyer's market for feature films and sports, the principal ingredients in the repertoire of offerings advertisers buy to attract would-be cus-

---

[22] Although television's labor force apparently belongs to a very large number of separate and sometimes independent unions, the major unions involved seem to be (1) American Federation of Television and Radio Artists (AFTRA: photographed employees); (2) National Association of Broadcast Employees and Technicians; (3) International Brotherhood of Electrical Workers (both for television cameramen); and (4) International Alliance of Theatrical Stage Employees (film cameramen).

tomers. If the networks were confronted with a pay-cable industry to compete against for these programs, they would probably have to pay more for them.

National television advertisers also gain. Demand for national television advertising traditionally has been very inelastic,[23] with the networks flowing costs straight through to advertisers.[24] Reasons for this inelasticity include the fact that there are no realistic alternative national advertising media—*Life* and *Look*, for example, are long gone—and the firms that predominantly purchase network advertising time, generally speaking, are leading firms in highly oligopolized industries.[25] Lower network programming costs mean lower advertising costs to network advertisers. Lower programming costs have the additional value to the networks of enabling them to hold onto major customers, who are certainly sophisticated enough to realize that the commodity they are buying, while relatively inefficient, nonetheless yields sufficient benefits considering its relative low cost.

National advertising, such as the networks sell for the most part, is inefficient to the extent that Florida viewers are not prime Prestone customers, for example, nor are Chicago residents the most likely buyers of suntan lotions. Traditionally, the transaction costs asso-

---

[23] Experience following the FCC's "imposition" of the "Prime Time Access Rule" is reasonably indicative of this inelasticity. The rule cut back the number of prime-time hours of programming the networks could feed local affiliates by one hour—the equivalent of twelve minutes of commercials. Network revenues simply increased about $500,000 per night. A scarce commodity had simply been made scarcer, and so the price was hiked.

[24] "Demographics" and other variables notwithstanding, networks, indeed most television, charge basically at the rate of $2.50 per thousand households per 30-second commercial (prime time). Total charges are keyed to Nielsen ratings, each point in the ratings being equivalent to about 700,000 households nationwide. For a very high rated program drawing a "30" in the ratings, charges would be, for example, 30 times 700,000 times $2.50 divided by 1,000, or $52,500 per 30-second commercial, or slightly more than five cents per household. There are 12 minutes of commercials allowed under the NAB Code per prime-time hour; at other times, 14 minutes or more of commercials are aired, for an average total per station of 190 minutes a day—the greatest percentage of any television system in the world (and, in minutes, roughly twice the daily output of "news, public affairs" programming).

[25] In 1970, for example, twenty-five advertisers accounted for about one-half of *total* television advertising expenditures of $2.8 billion. The top five firms alone accounted for about 20 percent of this total. For an analysis of the relationship of television advertising expenditures and concentration in the consumer goods industries, over the period 1947-1967, see *Industrial Structure and Competition Policy*, Staff Study Paper No. 2 for the Cabinet Committee on Price Stability, January 1969, pp. 31-32. See generally U.S. Congress, House, Small Business Committee, *Hearings on Advertising and Small Business*, 92d Congress, 1st session (1971).

ciated with buying time on a plethora of stations in a number of markets exceeded likely gains, considering the alternative of simply dealing with the networks. The availability of computer technology has changed things; so has increasing emphasis on local advertising, reflected by the fact that "local" and "national spot" components in the total advertising budget are becoming increasingly important.[26] "Cheap" national rates enable the networks to hold onto their major customers and thus at least control the rate at which change may occur in the pattern of national advertising.

The movie industry loses under the FCC pay-cable rules, since in effect they are compelled to grant the networks the equivalent of a right of first refusal to buy their product at the rates the networks are willing to pay. Indeed, the movie industry loses under the pay-cable rules any opportunity to sell any film that is more than three and fewer than ten years old, whether it is "suitable for television" or not. The movie industry is not even accorded a guaranteed market, moreover, since the FCC does not now, and is unlikely to, require the networks to purchase and air movies to any extent. The FCC rules thus have the effect of exacerbating the anticompetitive features of the network monopsony which is being challenged in the department's present suit against the networks.[27]

The theater industry at least potentially loses under the FCC rules to the extent, if any, that the pay-cable industry may erode the three-year-after-release period, during which theaters previously enjoyed some freedom from competing would-be buyers. As a practical matter, this does not apparently occur. As developed during the hearings of the Hart Antitrust and Monopoly Subcommittee in the summer of 1975, it is the networks which are eroding the three-year period. They are purchasing exclusivity against any cable showing of a film

---

[26] "National spot" is advertising sponsored by national firms and "spotted" on a preselected basis in certain markets. A further reason for its growth is the decision of the Federal Trade Commission that advertisers may not be compelled to purchase television advertising coverage they do not want. So long as the price stays low, advertisers are unlikely to balk at buying the basic sixty-station complement which is required for the most part by the networks.

[27] The sheer dollars involved in purchasing the products which are being contested—an average $800,000 per movie, for example—tend to indicate that it is not the typical television station whose right to purchase is being protected. Local affiliates, of course, have an obvious vested interest in the network's fare. Indeed, to the extent it is true that "local licensees have the mentality of theater operators" as Walter Cronkite has remarked, maintaining the status quo in and of itself provides an incentive for them to support the network position against pay-cable competition.

very far in advance of the date when the movie may be available for other than theatrical exhibition.[28]

The cable television industry loses something as a result of the FCC pay-cable rules, but it is unclear if their losses are as great as claimed. It is not clear, for example, that these rules deter the wiring of the major markets so much as general economic conditions and general marketing realities deter such activity. A complete rescission of the pay-cable rules, therefore, does not necessarily mean that the nation will be wired tomorrow. Nor would lifting the rules necessarily restore the cable television industry to the levels of prosperity once encountered, since only a small fraction of revenues from pay operations generally is obtained by the cable system itself. Also, there is relatively little indication that pay-cable services necessarily are a decisive advantage in collecting and retaining cable subscribers.

The professional sports leagues also lose something because of the FCC pay-cable rules. From a "clean hands" standpoint, however, the equities of their case are not altogether compelling. Some of the professional leagues, most notably professional football, substantially benefit from a wide range of antitrust exemptions (and favorable tax treatment) ostensibly justified as promoting the public's access to and enjoyment of sports. The antitrust exemption that allows the professional football clubs to sell their television rights as a block were indeed solicited from Congress on the assertion that this would result in expanded, more diverse television coverage. So long as the professional sports interests continue to try to constrict the supply of their product, and the public's access to it, a case can be made that "siphoning," as the FCC perceives it, could actually occur.

The public is the clearest loser under the FCC pay-cable rules—most clearly the 12 percent of the populace which is presently "on the cable." The rules constitute a federal policy determination that advertiser-selected programming is to be fostered to the effective exclusion of alternative ways of providing that programming. This FCC value judgment is imposed on all households, without regard to economic or technical considerations and, indeed, without regard to facts of any kind to begin with.[29]

---

[28] The extent to which the networks will go to preclude any competition is demonstrated, for example, by one National Broadcasting Company (NBC) purchase of a number of Metro-Goldwyn-Mayer (MGM) existing and planned films. The purchase was conditioned upon the grant of exclusivity against any pay-cable use, and even any use by way of video cassettes or video discs.

[29] This is, in a sense, "price discrimination" as one Office of Management and Budget (OMB) official has stated. It is different from other forms of regulatory price discrimination, since the FCC is not requiring cable subscribers to pay

Agency Positions. Three agencies have stated clear positions with respect to the FCC pay-cable rules. These are the Office of Tele-communications Policy, the Department of Justice, and the FCC, of course. The Department of Commerce has on occasion argued against continuation of the existing rules, but it has not proposed any firm position.

Office of Telecommunications Policy. OTP has strongly recommended rescission of the movies rules, by legislation if necessary. OTP argues, in essence, traditional policies support the general proposition that parties seeking government protection should bear the burden of proof about the need for such special treatment. The FCC has reversed the burden of proof and requires the pay-cable industry to prove that it will not cause any harm to existing broadcasters. OTP also argues that the test should be whether harm results to the public, and not whether any given industry group may suffer.

OTP recommends that the sports rules be relaxed, but it has not recommended that they be dropped in their entirety. OTP has suggested that the rules be a feature of proposed legislation with special protection included in such proposed legislation for name events, like the Super Bowl.

Department of Justice. The Department of Justice has consistently opposed all features of the existing pay-cable rules since they were adopted in 1970. It argues that the rules are nothing more than an effort to prefer one form of communication over another, and to favor one form of funding programming over another. In its most recent comments to the FCC, the department asserted that there were no facts in the record upon which the FCC would base such sweeping constraints. The department concedes, however, that the FCC can adopt restrictions when and if it finds factual evidence that "siphoning" has occurred *and* that that siphoning has resulted in actual harm to the public.

The department has consistently recommended that the rules be repealed altogether, through legislation if necessary.

Federal Communications Commission. The FCC asserts that the pay-cable rules represent a reasonable compromise between the extreme positions of the cable and the broadcast industries. At one time, the FCC position was that cable should concentrate on providing the kinds of programming that the existing mass appeal system of

---

more, except in the sense of an opportunity cost. Rather, the FCC prohibits them from paying more—indeed, from paying anything—for the reason that otherwise there is the "mere potential" others might lose what they already have.

broadcasting does not provide. If cable were to do that, the FCC argues, it would not incur any regulatory barriers; it is only when cable seeks to provide what the television industry now provides that the commission must act. The present chairman of the FCC repeatedly has stated that the rules as adopted were compelled by congressional pressures. By the same token, it is maintained that the rules give cable television much more than the Congress would give it.

The FCC is in the process of devising its own legislation, essentially to codify its present regulations in most regards.

Department of Commerce. At the time when the executive branch first undertook to study the feasibility and desirability of cable television legislation, the Commerce Department advanced classical policies that government ought not undertake to prevent people from buying what they want, unless the government has established on the record some overriding public purpose that would be compromised.

Commerce at that time argued a "hands off" posture towards the cable television industry, arguing that such a posture would best promote its development or, more specifically, best assure its development if the public wanted cable services. The recent report of the Committee on Economic Development suggested that serious thought be given to granting regulatory authority over cable to an agency other than the FCC, and inferentially at least proposed the Department of Commerce for that purpose.

Options

Outright Repeal. Propose outright repeal of FCC restrictions on both movies and sports programming for pay, and prohibit the FCC from adopting any pay programming prohibitions in the future.

Pro arguments.
- Would be consistent with the so-called free enterprise spirit— that is, government should not tell the public what they cannot buy "for their own protection."
- Solution like this is most immune to subsequent regulatory tampering.
- If "siphoning" occurs and it injures the public, Congress can amend the statute.

Con arguments.
- Easily criticized as ill-conceived, poorly thought-out effort to short circuit regulatory efforts aimed at protecting public interest.

165

- No guarantee "pay television" will not develop; if the networks ever thought they could make more money via pay cable, the public would really end up paying for what they now get for free, and would get commercials to boot.
- Overtly favors the cable industry.
- The usual arguments that competitive market forces can substitute for the dead hand of regulation do not necessarily work; sports programs are controlled by a government-sanctioned cartel, and movies are produced by an oligopolized industry.
- Would trigger massive broadcast industry resistance and opposition to any cable legislation. Congress, dependent on the broadcast industry because of recent campaign reforms, might react by adopting pay-television rules much stricter than anything FCC has ever come up with.

Caveated Repeal. Propose the outright repeal of FCC restrictions on both movies and sports, with the caveat that, if the FCC finds that siphoning has actually occurred to the detriment of the public interest, they can impose restrictions.

Pro arguments.

- Would be consistent with traditional policies that government should not prohibit all instances of potentially beneficial private behavior just because some instances may injure some people. FCC has more than adequate authorities to safeguard the public interest, if a clear need is demonstrated.
- Would probably result in no restrictions with respect to existing cable systems, most of which serve areas where broadcast service is technically deficient.
- Would avoid, somewhat, the appearance of prejudging the "siphoning" rationale; would possibly be more palatable to a Congress skeptical of the "good intentions" of either the cable or the television firms.

Con arguments.

- The FCC could very easily just make the requisite "findings" in a rule making. "Findings of fact" by a regulatory agency are technically not reviewable, and are certainly much more difficult to reverse on appeal than the present predictive judgments. FCC would just make findings and return to the status quo ante; "de-regulation" would amount to little more than "dynamic inaction."

- Any market-by-market standard seems to entail regional differentiation—different protections for different parts of the country, and different treatment of the population; the 1934 act was supposed to encourage equal service for everyone so far as possible.

Outright Repeal, Movies. Propose the repeal of the movies restrictions only, and prohibit any regulations in that regard in the future. Leave sports topic as is.

Pro arguments.
- Keeps sports out of the matter, in theory. No great loss since the present sports rules (for example, the "highwater mark rule") are so capricious in some ways they are very likely to be upset in the pending pay-cable appeal.
- In theory, the supply of movies is elastic; no problem of a "static pie." Increased demand for movies may lead to greater diversity in current studio output. Would at least loosen networks' present grip on the program production business.
- Does not run up against any strong public feeling; unlike the case with sports, the American public does not feel they have a constitutional right to see movies on television.
- Present FCC restrictions amount to "double hop" ancillarity—that is, efforts to regulate the marketing practices of the studios. No historical support for the proposition that FCC should act to assure the availability of the ingredients to broadcast service.

Con arguments.
- Concedes the validity of the siphoning rationale, at least in part; so, lends validity to regulatory status quo.
- If the movie makers really want to sell to pay television, they can do so—all they have to do is trim back the time for theatrical exhibition. What they really want is the right to extract every penny from the public.
- Price of films will be bid up. Television will pay more, so Procter and Gamble will pay more, so consumers will pay more. Plus that, a lot of consumers will pay twice—once to the cable, another time at the store.

Caveated Repeal, Movies. Same as caveated repeal above, but FCC could impose restrictions upon a finding that there had been an adverse impact on the availability of movies on television to the detriment of the public.

Pro argument.

- Basically like the caveated repeal above.

Con arguments.

- Legitimizes FCC regulation of the marketing practices of the movie industry; potentially expands regulation.
- Involves the FCC legitimately in weighing the public interest quotient of particular movies.

Outright Repeal of the Movies Rule, Caveated Repeal for Sports.

Pro arguments.

- Repeal of the movies rules is most justifiable. No valid purpose behind FCC's regulation of movie availability.
- Special treatment for sports arguably justified by oligopolized nature of that business.

Con arguments.

- Effect of movies repeal might be delayed because networks have bought exclusivity with respect to many of the best movies that now exist and are ripe for other than theatrical exhibition. Cannot constitutionally take those exclusive rights away without at least contemplating some kind of compensation.
- Putting movies beyond FCC control will trigger the "obscenity" assertions raised by some broadcasters before the FCC.
- Questionable whether the mass of the public feels strongly one way or another about being able to see movies at home.
- Will trigger massive broadcaster opposition.

Do Nothing.

Pro arguments.

- Possibly the most politically appealing route.
- District of Columbia Circuit Court may very well reverse the present pay-cable rules. If that happens in whole or in part, it may be easier to block either legislation or regulatory action prompted by broadcasters.
- Maintenance of status quo with respect to two most politically sensitive issues may avoid "poisoning the congressional atmosphere against less visible yet important proposals to deregulate cable."

Con arguments.

- Cable legislation not addressing major issues does not require presidential attention.
- Past efforts would be seen as "the elephant laboring and producing a mouse." Alternatively, would seem as if the television interests had succeeded in intimidating the executive branch as well.
- Congressional staffs are prepared to tackle the issues. Weak proposal by the administration will be seen as tantamount to a judgment that that is all that is appropriate.

Cover and book design: Pat Taylor